THE PSYCHOLOGY OF TIME

THE PSYCHOLOGY
OF TIME

PAUL FRAISSE

Professeur de Psychologie Expérimentale à la Sorbonne
Directeur à l'Ecole Pratique des Hautes Etudes
Directeur de l'Institut de Psychologie
de l'Université de Paris

TRANSLATED BY JENNIFER LEITH, M.A. (OXON.)

GREENWOOD PRESS, PUBLISHERS
WESTPORT, CONNECTICUT

Library of Congress Cataloging in Publication Data

Fraisse, Paul.
    The psychology of time.

    Translation of Psychologie du temps.
    Reprint of the ed. published by Harper & Row,
New York.
    Bibliography: p.
    Includes index.
    1. Time--Psychological aspects. I. Title.
[BF468.F6813   1975]   153.7'53   75-37653
ISBN 0-8371-8556-4

Originally published in 1963 by Harper & Row, Publishers,
New York

Reprinted with the permission of Harper & Row, Publishers, Inc.

Reprinted in 1975 by Greenwood Press, Inc.,
51 Riverside Avenue, Westport, Conn. 06880

Library of Congress catalog card number 75-37653

ISBN 0-8371-8556-4

Printed in the United States of America

10 9 8 7 6 5 4 3 2

CONTENTS

TO MY WIFE

||||||||||||||||||||||||||||||||||||

INTRODUCTION

MAN LIVES AMIDST CHANGE. EVEN BEFORE HE BECOMES AWARE OF the fact that he himself is changing, he sees continuous changes in the world around him. Night succeeds day, good weather follows bad, winter comes after summer. Animals are born and die; nothing can stop the flow of the river or the erosion of the rock. Everything is caught up in this change, even man. His biological, psychological, and social life consists entirely of change.

But unlike all other forms of life, man knows that he lives amidst change. He can reconstitute past changes through memory and thus establish laws which help him foresee the future chain of events. He learns very early in life to make use of change instead of merely submitting to it.

The birth of the notion of time is no doubt the result of the experience of successions, of which some are periodic and others not, of continuous and discontinuous changes, of interwoven renewals and relatively permanent states. This experience may even explain the word itself. Even in highly developed languages, the word *time* is commonly used to denote the moments of change: "to do something all in good time," "he is not of his time," "for all time." In French the word *temps* is used in the more concrete sense of "weather," or the successive states of the atmosphere.

1

The Latin word *tempus,* from which it is derived, also has this double meaning. The Sanskrit root of the word meant to light or to burn; this points to the fundamental nature of our experience of the rhythms of day and night. The word *day* is still used poetically in the sense of light (Regnaud, 1885). As far back as we can trace, therefore, the word has had both a concrete and an abstract meaning, and this is still so today.

Through the centuries man has striven to master the fundamental conditions of his existence. Periodic changes—day and night, the lunar cycle, the annual recurrence of the seasons—have provided at the same time natural frames of reference against which to locate other changes, and a means of measurement. For thousands of years scientists have been studying these periodic phenomena and trying to relate them to each other. Their work is not yet done. We are still perfecting methods of measuring time, and the reform of the calendar is on the United Nations agenda. The anguish suffered by men aware of the changes taking place within them and their inevitable end has caused sages and moralists to study the actual significance of change for individuals, societies, and the world. Starting out from an increasingly abstract notion of time, philosophers have reflected on its nature. The history of time is inseparable from the history of human thought.

Let us consider how the philosophers of the western world have treated the problem. We know that they have not been much concerned with the origin of the concept of time nor with its nature, in so far as this is also an idea; they have kept closer to reality. They have sought the relationship of time and its manifestations to movement. Is this eternal or not? Does it exist independently of a mind which links the before and the after? These questions are not fully answered. They are reborn with every generation, like those which face the moralist, with which they are closely linked. According to Platonic philosophy, time is the mobile image of eternity, revealing itself in a world governed by cycles of recurring change. All Judeo-Christian thought is shaped by the revelation of a world created with its own time, the background for the story of sin and redemption; beyond death, in the Kingdom of God, time returns to eternity. The modern world has discovered the unlimited antiquity of its history; the doctrine of evolution is established and has given rise to the idea of a sort of "immanence" of time; thus time becomes the scene of the indefinite progress of man.

Descartes' ideas began a critical phase in the history of philosophy which faced man with questions of a different kind. Where do we get this *notion* of time, and how is it related to our immediate experiences? This was a question of epistemology which was to lead to more specifically psychological questions. Of course man in general and philosophers and moralists in particular have always been concerned with psychological problems. Their works are full of facts they have experienced and a historian can easily see how their philosophical ideas correspond to their way of experiencing time. But as soon as their thought is centered on the origin and significance of the *idea* of time, their perspectives turn from meditation on God and the world to man, and especially to the laws which govern his mind.

All the philosophers, including Kant, who have sought the origin of our idea of time, have agreed that it comes from change. Aristotle noted "that time . . . does not exist without changes." (*Physics*, book IV.) But what changes? The changes in our sensations or the changes in our thoughts? The answer to this question depends on every philosopher's individual conception of the idea.

Condillac held a theory of integral empiricism. His statue "would have known no more than an instant if the first odoriferous body to come along had acted on it consistently for an hour, a day or more. . . . Only a succession of smells sensed by the organ or recalled by the memory can give it some idea of duration." (*Traité des sensations*, 1921, p. 85.)[1] Hume thought along the same lines. "A man in a sound sleep, or strongly occupy'd with one thought, is insensible of time. . . . Whenever we have no successive perceptions, we have no notion of time, even tho' there be a real succession in the objects . . . time cannot make its appearance to the mind, either alone, or attended with a steady un-

---

[1] We have borrowed this quotation, and several others to follow, from Sivadjian's *Le temps*, 1938, in which a considerable number of texts on time are examined.

changeable object, but is always discover'd by some *perceivable* succession of changeable objects." (*A treatise on human nature,* vol. I, p. 342.)

Descartes, on the other hand, believes that our notion of time has its origin in our inner experience; he does not discriminate between the idea of time and that of duration. ". . . When I think that I am now and also remember having been in the past, and I conceive several different thoughts, whose number I know, I acquire within me ideas of duration and of number which I can shortly afterwards transpose to anything else I like." (*Troisième méditation, Œuvres,* vol. I, p. 66.) In the same way Locke says: "For whilst we are thinking, or whilst we receive successively several ideas in our minds, we know that we do exist; and so we call the existence, or the continuation of the existence of ourselves, or anything else, commensurate to the succession of any ideas in our minds, the *duration* of ourselves, or any such other thing co-existent with our thinking. . . . But wherever a man is, with all things at rest about him, without perceiving any motion at all,—if during this hour of quiet he has been thinking, he will perceive the various ideas of his own thoughts in his own mind, appearing one after another, and thereby observe and find succession where he could observe no motion." (*Essay concerning human understanding,* vol. I, pp. 239, 242.)

This idea of time is born of the life of thought; the empiricists wondered how it could apply to the outside world. Hume formulates the problem: "Ideas always represent the objects or impressions, from which they are deriv'd, and can never without a fiction represent or be apply'd to any other." (*Op. cit.,* p. 344.) Condillac scoffs: "You apply your own duration to everything outside you and by this means you imagine a common measurement which is commensurable, instant for instant, with the duration of everything in existence. Are you not thereby realizing an abstraction?" (*De l'art de penser,* p. 149.) On the other hand, an entirely empiricist viewpoint, like that represented by Condillac and Hume, leads to extreme relativism, for in their opinion the idea is nothing but the double of our sense experience.

Kant, feeling the necessity for a common time at the root of the laws of science, postulated that there could only be one pure form of intuition of the senses. In his opinion, the unity of time cannot arise from a diversity of sensations but only from the way in which this diversity is linked together by the mind. Kant's critique opposes all attempts to attain an absolute world time or ego time. But it is obviously wrong to think that he believed in the innateness of the notion of time. Being a notion, this is "acquired, being drawn not from some sensation of objects (for sensations provide the substance but not the form of human knowledge) but from the operations of the mind itself in accordance with the constant law governing the sensations of the mind." (Kant, *Dissertation of 1772*, quoted by Sivadjian, *op. cit.*, p. 164.) What is innate is the possibility of forming representations of different sensations in the form of temporal relationships. The notion of time is ideal because it is not abstracted from experience, but it only becomes apparent through the activity of the subject. This is at least the interpretation suggested by Havet, the most recent commentator on Kant. The latter made little attempt to explain the actual processes of the birth of our notion of time; this was not his aim.

Kant showed that our notion of time is not a copy of things but a way of considering them. He thus prepared the way for the psychologists by diverting them from the search for a reality "in itself," and indicating the origin of the notion of time in the activity of the mind which thinks and relates various changes. The *Critique* also had an indirect influence on subsequent thought. By making time a form of sensitivity, he shifted the whole problem. After him, philosophers, and later psychologists, were less concerned with the notion of time than with our awareness of time.

In the post-Kantian period, the problem has gradually changed from the plane of knowledge to that of psychology. It is no longer so important to know under what conditions science can be founded on the notion of time; there is an increasing tendency to seek the empirical origin of our notion of time and, more generally, of our awareness of the two fundamental aspects of this notion, succession and duration. Kant's ideas are everywhere apparent behind this trend, but particularly in Germany. Authors often confuse the metaphysical a priori reasoning, which is truly

Kantian, with a doctrine of innate ideas; they amass arguments against the hypothesis that time is an a priori intuition. But even for them one point seems to be clear: the succession of sensations or thoughts is not enough on its own to give the notion of succession. This can only arise from the apprehension of a relationship. From the nineteenth century on, this relationship was sought among the different representations that may be formed of reality: besides perceptions, which alone could not give us the elements of a succession since each one takes place in an instant, we also have the images provided by our memory. These reproduce the sequence of experienced events through the laws of association, and we thus become aware of the relationships of the before and after which link them. The very fact that memory and association are constantly referred to is a clear enough indication that the problem was now being considered more and more from the psychological point of view.[2]

The relationship between images and sensations or between several images was obviously viewed in different ways by the theorists of this period. Herbart, for instance, considers that if, after a series of representations a, b, c, d, e, the first element a is presented again to the consciousness, it calls to mind the other elements b, c, d, e, which were associated with it. There is, therefore, a presentation of succession, or in other words a change due to a process of evolution. If, on the other hand, e is presented again, there is a *process of involution*, through which d, then c, then b, etc., are called to mind, but the further each element is from the first, the less clear it is; in this seriation there is not the same clarity of presentation. Each process, therefore, consists in the simultaneous presentation of the two terminal elements of the series and a combination of the two processes results in a complete perception of time (according to Nichols, 1890).

For Spencer, time is not possible without the establishment of a relationship between states of consciousness. Awareness of time has its origin in consciousness of the different positions of successive impressions in relation to the present impression; this difference in position is entirely due to our consciousness of the fact that these impressions do not all exist together. But the perception of these relationships of position is only the raw material from which we go on to construct a notion of time. This is ultimately the notion of a relationship of positions but it is dissociated from any particular position.

According to Wundt, the simple repetition of a sound is enough to provide all the elements for the perception of time. When the second sound is produced, it somehow reproduces the first whose image is still present. The recalling of the first sound by the second thus provides the beginning, the perception of the second —the end, and the persistence of the image—the length of the interval. A series of conscious facts thus leads to a temporal relationship because there is always a great number of durable representations among them. Time has its origin at the same time in this succession and in the relative simultaneity of the psychic formations.

Guyau also represents this school of thought but his conception is more dynamic. His aim is to find the elements of the experience of time. He establishes two: the bed of time, formed by the succession of our representations which tend to become dim with distance; and the flow of time, or the perspective introduced between our representations by desire and effort. Consciousness of time has its origin in the association between varied images of different degree and inner facts connected with affectivity.

The one point which *all* the authors of this period have in common is their attempt to explain our notion of time by analyzing our states of consciousness. However paradoxical it may seem, Bergson's approach is in this respect essentially the same although his line of thought is more metaphysical than psychological. He also refers to our inner experience, but instead of finding multiplicity here, he sees the intuitive unity of the uniform duration of the ego, in which there is an intimate fusion of states which only seem to us to be successive because our successive sensations "retain something of the mutual exteriority which objectively characterizes their causes." (*Essai sur les données immédiates de la conscience,* 19th ed., 1920, p. 95.) This line of thought is psychological in its analysis of the problem and philosophical in its aims; it is still valid even now, for it can be regarded as a permament attitude of the human mind.[3]

Around the middle of the nineteenth century there evolved an entirely new approach to the problem of time: the empirical study of the accuracy with which men perceive time. Under the influence of psychophysics, using the methods recently described by Fechner, the psychology of time moved to the laboratory. The first experimentalists tackled the classical problems of psychophysics: does Weber's law apply to time? Are there constant errors in the perception of time? What influence do the contents of temporal intervals exert on perceived duration, etc.? There was an increasing amount of research carried out in Germany (Mach, 1865; Vierordt, 1868; Kollert, 1883; Estel, 1884; Mehner, 1883; Glass, 1887; Ejner, 1889; Münsterberg, 1889; Meumann, 1893–1896; Schumann, 1898). At first, the research took place on two different levels. On the one hand, by means of experiments, the psychologists sought to determine what the subject perceives by studying what he does (e.g., reproduction) or what he says (comparison). On the other hand, like the philosophers of their time, they tried to determine the foundations of consciousness of time by methods of introspection. The new element was that the experimentalists were no longer content with their own personal observations but used those of their "subjects," usually their assistants or colleagues.

Gradually these two approaches separated. The introspective work, aiming at the establishment of first contents and immediate experiences, petered out in abstractions. Despite the advent of the Würzburg school, bringing systematization of introspection,

[3] It is now mainly represented by the school of phenomenology whose followers start out from our experience and try, through a transcendental reduction, to strip it of all that is fortuitous and retain only its essential significance. The phenomenologists have in fact been much concerned with the analysis of time. Husserl (1928), Heidegger (1927), Merleau-Ponty (1945), and Berger (1950) stressed that time is not an object and can therefore be neither a given fact nor a container or a content. For them the essential factor is the temporality of the conscious, revealed by our only experience, that of the present; the present does not exist without its horizons, for it is the present of a being in a perpetual state of change. Consciousness reveals time which thus seems like a dimension of our being.

its followers found that the essence of perception escapes introspection. On the other hand, the results obtained by the
experimentalists were becoming more and more coherent and rewarding. The psychologists were no longer content to study the
perception of time in normal adults, and were extending their
research to animals, children, and the mentally ill. In these
cases they could study perception through methods of conditioning or even through verbal responses, but the subjects could not
describe their conscious experience. The attitude which had been
implicitly present since the very first experiments gradually became more apparent: it was essential to study what man *does* in
reaction to the situations in which he finds himself.

Despite the "simplism" of the first behaviorists, the whole of
psychology turned into a science of human behavior in the first
quarter of the twentieth century. The attitude of this new psychology to time was formulated in the paper read by Henri
Piéron at the International Psychological Congress at Oxford in
1923 and in the course of lectures given by Pierre Janet in 1927–
1928 at the Collège de France on *L'évolution de la mémoire et
de la notion du temps*. Piéron was only concerned with an exploration of the psychophysiological problems of the perception
of time, but this led him to define a general method: these problems must be tackled "on the objective ground of an analysis of
human behavior in relation to time" (1923, p. 1.).[4]

With his usual originality, Janet reshaped the perspectives of
the study of time. In his first lecture he stressed that psychology
has better things to do than to concentrate on the study of
thought; it must start out from action. The only question we
should ask ourselves is "What action do we have on time?" According to him, the first action relating to time is behavior of
effort which, like expectant behavior, gives rise to feelings of duration. This feeling is not a primary action but a regulation of
action due to the necessity of our adapting to irreversible changes.
When we meet someone who looks older than when we last saw
him, we become aware of the fact that time has elapsed between

[4] This orientation produced good results, not only in the study of problems
pertaining to perception but also in that of our most personal attitudes;
Piéron later observed this himself in another paper read to the *Association
française pour l'Avancement des Sciences* in 1945.

the two meetings. The notion of a universal, homogeneous time in which all changes are located is in itself the result of a social type of behavior; it is the necessary framework within which uniformity is bestowed on all those individual durations which are basically heterogeneous.

This is not the moment to discuss Janet's theory in detail; we would only stress the originality of his aims. The psychological problem is no longer to know either what time is or what is the nature of our notion of time, nor is it even to seek the genesis of time in some intuition or construction of the mind; it is to understand how man reacts to the situation imposed on him of living in time. Far from being misconstrued, the data supplied to our conscious mind find their true significance in this situation. They are not a mere copy of reality but "a collection of signs, formulae and useful interpretations" (Wallon, 1930, p. 326), which develop in action itself and serve in return as a guide for our activity as we become conscious of them.

This is also the perspective of our work. We shall study the different ways in which man adapts to the temporal conditions of his existence; this is what we call "temporally organized behavior."

Basically these temporal conditions all stem from the fact that we live in physical, technical, and social backgrounds which are changing all the time. We do not only undergo these changes, we create them, for our own activity is itself nothing but a succession of changes.

Without prejudice to their ultimate significance, it seems that all changes, whether continuous or discontinuous, periodic or not, have two aspects. Where there is change, there is a *succession* of phases of a single process or of various concomitant processes. In its turn, succession implies the existence of intervals between the successive steps. These intervals vary in length; we say that they are more or less durable, according to what remains relatively unchanged in them. We speak of the duration of a day, meaning the period of light which stretches from the end of one night to the beginning of the next.

Obviously these successive phases and intervals are relative to the content of the change and to the aspect on which we choose to fix our attention. A day is the interval between two nights, but within this day we can discern changes where there are again successions and intervals. It is only important to note that this double nature is evident, whatever the phenomenon observed and the scale on which it is seen.

We react in very different ways to these temporal conditions. It is possible to differentiate between three main groups of reactions which correspond to three levels of adaptation: (1) conditioning to change; (2) the perception of change; and (3) control over change.[5]

1. Conditioning to Time

The first level of adaptation is biological and is common to animals and man. Whenever the changes to which we are exposed have some measure of regularity, they give rise, through conditioning, to synchronous changes in our organism.

If the changes are periodic—the most important of these for man being the day-night cycle—they induce activities in our organism which have the same rhythm. The regulation of this rhythm is at first exogenous, but it gradually becomes endogenous and relatively independent of the environment. This has the effect of bringing our lives into harmony with the principal changes going on around us. Furthermore, the periodicity of the modifications taking place in the organism provides it with a physiological clock which is used by man as well as animals for temporal orientation, especially when the cues usually afforded by the changes in their environment are lacking (Chapter 1).

These acquired physiological structures are also apparent in trace conditioning through which animals adapt to a regular interval between two or more changes. Instrumental conditioning also shows that they are capable of learning the practical significance of an interval.

This registering of duration on the level of biological reactions

[5] Our reactions to changes in ourselves over a long period, i.e., to the different ages of life, are outside the scope of this book. A study of this subject would raise another whole set of problems. There is in fact a psychology of childhood, of adolescence, of maturity, and of old age, which deal with the reactions peculiar to each age in the face of change. We shall, however, make a systematic study of the way in which we adapt to the changes in our environment at every stage of life.

also plays a role for man, but in his case it is usually obscured by conscious evaluations of the duration (Chapter 2).

2. The Perception of Time

Within temporal limits which are of course narrow but still of great practical importance, we can perceive change. This perception is characterized by the integration of successive stimuli in such a way that they can be perceived with relative simultaneity; rhythms and sentences in speech are obvious examples. This simultaneity defines the psychological present, within which we perceive the fundamental characteristics of change: the order of the stimuli and the intervals between them (Chapter 3).

Under what conditions can we pass, within the psychological present, from the perception of the instantaneous to that of the durable and from the simultaneous to the successive? This question is discussed in Chapter 4.

What are the modalities of our perception of duration? These cannot be defined without studying the relationship of the perceived duration to the nature of what is changing and the structure of the processes of integration of the succession (Chapter 5).

3. Control over Time

Perception only permits us to apprehend changes at the moment of their occurrence. Man is not restricted by this limitation because he can form representations of these changes; he can thus relate them to himself and to each other, and make use of them to some extent for his own ends.

Through memory we can reconstitute the succession of experienced changes and anticipate changes to come. By these means we acquire a past and a future, a temporal horizon in relation to which our every action takes on its full significance (Chapter 6).

We become conscious of duration through our feelings of time, which are fundamentally the awareness of an obstacle: the interval between what we are doing and what we should like to do in the near future. We evaluate this duration directly from the number of changes which we noticed taking place within it (Chapter 7).

The constitution of a temporal horizon and the appreciation of duration do not imply that we relate all the data concerning order and duration, for these remain to a great extent intuitive. The relating of such data takes place on a higher level through the intellectual operations which are at the root of our notion of time, the abstract thread linking all change. We can then measure time, reconstitute and make use of changes, without being bound by their apparent qualities or in particular by the irreversibility of experienced order (Chapter 8).

If we were to attempt a justification of our classification of time-structured behaviors at this point, we should be anticipating all that is to follow. On the other hand, the nature of the task we have set ourselves would be clearer if we were to relate our classification briefly to those most often suggested.

Let us first discard all those classifications which start out from different categories of time: physical time, biological time, psychological time, social time, etc. These describe the different series of changes but they are not made from the psychological point of view.

From the moment when psychologists first considered the problem of time, a fundamental distinction became apparent between the primary experience of duration, ascribed to a "time sense," and our rational idea of time. This distinction was reconsidered and transformed by Bergson, who saw it as the difference between time lived and time thought. Released from its metaphysical connotations, it is to be found, with different shades of meaning, in all psychological treatises, and especially in works on mental pathology (Straus, Minkowski, Ehrenwald).

The classifications based on this distinction are connected with a psychology which is concerned only with the data presented to the conscious mind. They contrast the different ways in which we think that we apprehend time. They are, therefore, inadequate for a psychology of behavior according to which, as we know, apprehension is only one—sometimes essential—part of an action. In our research we shall, of course, make use of the analyses prompted by the distinction between lived and thought time, but the standpoint from which we consider them, that of adaptation to change, will lend them a different significance.

Being limited to our states of consciousness, the above distinction does not take into account our biological adaptations and their psychological consequences. These adaptations were studied, however, by the psychophysiologists, such as Pavlov and Piéron, and figured in the classification proposed by the German neuropsychiatrist, Kleist (1934). Kleist based his classification on the diversity of nervous centers involved in pathological disorders relating to time and made a distinction between the following: (1) The registering of time, which is the basis for temporal orientation and depends on the nuclei and the vegetative centers of the hypothalamus; (2) The appreciation of length of time which may be connected with the activity of the vestibular centers; and (3) The apprehension of time structures, associated with the cortical centers.

This classification is very interesting and coincides with ours on several points. However, apart from the fact that the role of the vestibular centers in the appreciation of time has not been confirmed, it also has the disadvantage of not giving a clear enough definition of the behaviors which depend on the activity of the cerebral cortex, such as the perception of time structures, and the elaboration of the representation of time which cannot be specifically located.

The classification suggested by Delay (1942) combines the perspectives of the psychology of the conscious and those of neuropsychiatry. He distinguishes between: (1) sensory-motor time, or time as it is registered; (2) social time, uniform and measurable; and (3) autistic time, which is much the same as Bergson's psychological duration.

Our own classification is based on a similar idea, but it is more functional. We have aimed to differentiate between our various processes of adaptation to change; for this we have drawn on physiological facts, on pathology and genetic psychology as well as on an analysis of psychological functions. But, while using these various disciplines, we have remained faithful to the method of the psychology of behavior; through these disciplines we have tried to determine what man *does* in order to know time and use it, locating himself in the universal change that carries him through life.

||||||||||||||||||||||||||||||||||||||||||||||||||||||||||||||||||

PART I

CONDITIONING TO TIME

||||||||||||||||||||||||||||||||||||||||||||||||||||||||||||||||||||

The changes going on around us do not only cause immediate reactions to each of their phases. Their order and periodicity also induce, in the organism, series of physiological changes and modifications in behavior which have the same temporal characteristics. These series are such that once the first change has occurred, all the others follow in the same order and at the same intervals of time as when the original changes took place. In this case time acts as a conditioning stimulus.

Thus under the influence of periodic changes, the organism becomes a physiological clock which provides cues for temporal orientation both in animals and in man (Chapter 1).

The ability of the organism to reproduce regular sequences which have been undergone by, or created by the activity of, animals or humans permits them to appreciate duration; this can be seen in the case of trace or operant conditioning (Chapter 2).

The fact of conditioning to time explains why animals can adapt to change through temporally organized behavior. Man also makes use of these biological mechanisms for temporal orientation and for estimations of duration, but in his case they are part of more complex behavior in which the symbolic knowledge of change also has a role.

‖‖‖

ADAPTATION TO PERIODIC CHANGES

MOST OF THE CHANGES WHICH OCCUR IN NATURE ARE PERIODIC AND there is more often than not a relationship between this periodicity and the movement of the universe. The background of life is punctuated by the tides, the alternation of night and day, the lunar cycle, the seasons; and living organisms themselves present a number of periodic phenomena: the pulse frequency; the respiratory cycle; the rhythm of the digestive organs, of sleep; menstruation; the seasonal rhythms of vegetable life, of sexual activity, migrations; etc.

Certain of the periodic phenomena of organic life are endogenous and bear no relationship to the alternations of nature. This applies to the rhythm of cerebral waves, of heartbeats, and even of breathing. Other organic changes occur over a period which coincides with a natural phenomenon, but it has not been possible to find any relationship of cause and effect: for instance, the menstrual cycle in women has the same frequency as the lunar cycle. This does not however, exclude the possibility that what seems to us to be pure coincidence may be due to the action of an agent as yet undiscovered, or perhaps to the persistence of an effect which arose during evolution.

Finally, many of the periodic activities of organisms are induced by the periodic changes to which they are exposed: variations in light, temperature, humidity, etc., governed by the rhythms of the cosmos. Thus many animals are active by day and sleep by night. Some do the reverse. The whole of nature follows the cycle of the seasons, which are in turn determined by the relative positions of the earth and the sun. Most important is the fact that these cosmic rhythms not only cause reactive activities but also, in many cases, give rise to a true periodicity which becomes to some degree an integral part of the organism of living beings. The organism becomes capable of anticipating changes in its surroundings to such an extent that if the action of the inducing agent is removed, the induced rhythms continue for a certain length of time. Thus, by induction, the periodicity ceases to be exogenous and becomes endogenous.

To explain this phenomenon, let us take two thoroughly studied examples from animal psychology.

The small flat worms, called *Convoluta*, which form large dark-green patches on the wet sand at low tide, burrow down into the sand as soon as the motion of the rising tide begins. In the case of young worms raised in an aquarium, these geotropic reactions, which are alternately negative and positive in step with the rhythm of the tides, do not occur (Martin, 1900). If, on the other hand, worms are placed in an aquarium after being exposed for some time to the action of the tides, they continue for several days to burrow down into the sand and emerge from it as if this action were still present. (Gamble and Keeble, 1905.) The glow-worm (wingless female of the *Lampyris noctiluca*), which emits light at night, to attract the male, but not by day, continues to glow only at night for four or five days even when placed in permanent darkness. Gradually this alternation disappears and the insect emits a constant but weaker light (Piéron, 1925).

These periodic activities are not, therefore, simply reflex reactions to stimuli associated with the rhythm of the tide or the day-night cycle, for they persist for a while even if the direct cause is removed. On the other hand, they are not of endogenous origin, because they do cease gradually when the organism is no longer exposed to periodic changes. It therefore seems that *rhythmic persistence*, as it is called by Piéron (*L'évolution de la mémoire*, 1910), is an effect of experience and an adaptation to change by anticipation.[1] This is, therefore, a true case of temporally organized behavior, as defined in the Introduction. Such facts are often explained by some sort of "sense of time."

Before enlarging on all the consequences of this rhythmic persistence, we must consider whether these phenomena are not just a curiosity for naturalists. Their importance will be easy to see, not only in the case of animals but also of humans, and we shall seek an explanation for this.

I *The Variety of Periodic Adaptations*

Periodic changes are to be found even in the vegetable kingdom. Many flowers open at certain hours of the day. Observing this, Linnaeus planted "Flora's Clocks," each hour being indicated by a different flower. For instance, Convolvulus (bear's bind) opens at about 3:00 A.M., the white water-lily at 7:00 A.M., the marigold at 9:00 A.M., the pretty-by-night at 6:00 P.M., etc. (Bonnier quoted by Piéron, *ibid.*, p. 51.)

The leaves of leguminous plants lie in one position by day and another by night. This alternation, which has been studied for a long time, is innate. If a bean plant, for instance, is kept under constant conditions, and in particular under a constant light, the alternation persists. Bünning (1935) even found that, although the predominant rhythm for this species is 12–12, there are certain varieties which have a 23-hour rhythm and others a 26-hour rhythm; the period peculiar to each variety is constant and hereditary. The interesting fact from our point of view is that only the actual alternation seems to be endogenous and the synchronization with night and day is induced by the circadian rhythm itself. The rhythmic movement of the leaves may be reversed if artificial lighting is used at night and the plants are kept in the dark during the day. However, it is also possible to obtain rhythms of 6 hours (3–3), 12 hours (6–6), or even 36 hours (18–18) by the alternation of light and dark (Pfeffer, 1915, quoted by Piéron,

[1] This chapter owes much to the work of Piéron, whose interest in these phenomena continued throughout his career. See particularly Piéron, 1910, 1937, 1945.

1937). If the plant then returns to constant light, however, these acquired rhythms disappear and 24-hour periodicity (12–12) is reinstated.

More general experiments carried out by Darwin and Peitz (quoted by Piéron, 1910) have proved the possibility of inducing persistent rhythms in plants by the periodic action of light or weight.

In the case of animals the facts are even more striking and very widespread. There are hardly any species in which it has not been possible to find seasonal or diurnal rhythms.

Circadian rhythms in animals, unlike plants, are almost always acquired.[2] In certain cases the induced rhythm seems to be caused entirely by the rhythms of the environment in which the animal lives, but for the most part the only function of these conditions is to regulate the necessary alternation of periods of waking and sleeping, activity and rest, which no organism can escape without dying. Diurnal rhythms are characterized essentially by the synchronization of this alternation with the regular cycle of day and night. This induction begins at the birth of the animal. Thus the chicken embryo has no rhythmic activity and even a young chicken kept in constant light has no cycle of activity adjusted to day and night. On the other hand, if it is subjected to alternations of light and darkness its behavior will follow this rhythm (Hiebel and Kayser, 1949). Its behavior cannot, however, be made to follow any indiscriminate periodicity. It has been shown by a large number of experiments that it is easier to adapt animals raised in artificial conditions to a 24-hour rhythm than to any other, and a number of authors, including Kayser (1952), think that there is a hereditary predisposition to a 24-hour rhythm. This could also be confirmed by the fact that it is easier to reverse the diurnal temperature rhythm of a pigeon (Kayser, 1952) or the rhythm of activity of a white rat (Hunt and Schlosberg, 1939b) than to change the period. In certain cases, however, it has proved possible to obtain rhythms of a slightly longer

---

[2] Hoffmann (1955), however, claimed fairly recently to have observed 24-hour rhythms in lizards which had been kept in constant conditions since birth. The duration of the period differed slightly from that found in animals raised in their natural surroundings.

period. In the case of white rats, 16-hour periods of activity have been made to follow periods of rest of the same duration (Hunt and Schlosberg, 1939b).

Despite various exceptional cases of learning of this kind, it is certain that no animal escapes the domination of the circadian rhythm, except perhaps deep-sea fish; these fish have been found to sleep at intervals far exceeding 24 hours (Piéron, *Le problème physiologique du sommeil*, 1912).

These examples show how the periodicity of external changes can influence the timing of phases of activity which correspond, by their very nature, to a biological necessity. Apart from these general rhythms of activity, other activities exist, especially among insects, which always take place at the same time and which are entirely the result of acquired experience. If, for several successive days, bees find food at the same place and the same time, they will subsequently come back every day at the same time; this will continue for several days after the supply of food is stopped. This training may be carried out for several different times of day at once (Beling, 1929); it can even succeed if the food is put in two different places at an interval of several hours (Wahl, 1932).[3] This sense of time is also found in other species. Fish come every day at the same time to the same place at which they are fed (Braunschmid, 1930) and birds show an increase in activity just before their feeding time (Stein, 1951).

In all these examples, the activity begins at the same point or points in the day-night cycle, and the possibility that external cues play a part cannot be excluded. Although the latter are used in normal life, however, it has been proved that they are not essential. Bees, for example, have been trained to come for food every 21 hours; in this case no cues could have been used from the outside world. (Beling, 1929.) Renner's recent experiment (1955) is even more conclusive. He trained bees to come for food

---

[3] Grabensberger (1933) thought he had established the same facts for ants. In 1943, however, Reichle showed that the activity of ants looking for food is directly related to climatic conditions. Dobrzanski (1956) repeated all Grabensberger's experiments systematically and showed, after putting out food at the same time and the same place for several weeks, that there was the same density of ants at the time of training as at any other time of the day.

at a certain hour in an experimental room in Paris. They were then taken, between two feeding times, from Paris to New York. Here they were placed in an identical room and deprived of food. On the following days they came for food at the same time as in Paris, regardless of the difference in the time of day between Paris and New York. When the experiment was repeated taking the bees from New York to Paris, the same result was obtained. This proves that bees have an internal control which acts independently of external conditions.

This intrinsic control must be formed by periodic variations in the organism induced by the circadian rhythm. Attempts to train animals to follow a rhythm of more than 24 hours have failed (Stein, 1951), no doubt because they could not find any cues, either external or internal.

The existence of an internal clock has also been shown in studies concerning the "solar" orientation of arthropods and birds, a subject which has recently been critically reviewed by Medioni (1956). Numerous experiments have shown that insects, crustaceans, and birds are capable of moving in a given compass direction, guided only by the position of the sun. For this behavior to be constant, the animal must obviously take the time of day into account, to allow for differences in the position of the sun. Thus bees which have been trained in the afternoon to go westward for food, go in the same direction the next morning, even if their hive has been moved to entirely different surroundings during the night and their exit turned in another direction. (Von Frisch and Lindauer, 1954.) A starling can also be trained to seek food in feeding places to the east, whatever the time of day.

This regularity governed by a physiological clock is particularly apparent when a discrepancy is introduced between its effects and the position of the sun. If the cage of the starling, trained in daylight to fly east, is placed in a dark, underground room in the center of a circular tent of white canvas which diffuses the light from an electric light acting as an artificial sun, the bird will turn toward the east at the usual time, its direction making the same angle with the artificial sun as it did with the real sun. If the same experiment is repeated a few hours later, the light being in the same position, the bird will err toward the west. It has allowed for a difference in the position of the sun, as if this really had changed. As the artificial sun has remained in the same position, it makes a mistake (Kramer, 1952).

It is also possible to prove the existence of the internal clock by setting it wrongly. After starlings have been trained to fly in the direction of one of the cardinal points at any hour of the day, they are subjected to the constant influence of an artificial day. This consists of alternating light and darkness, giving a faithful reproduction of the diurnal rhythm but with a discrepancy of six hours from the sun. If, after several days, the starlings are subjected to orientation tests by real light of day, they will be found at 3:00 P.M., for instance, to fly in the correct direction for 9:00 A.M. This error is obviously the result of the false setting of their internal clock, which has adapted itself to the new rhythm (Hoffmann, 1954).

The existence and nature of these physiological clocks imply that organisms are capable of adapting themselves to regular changes.

Experiments carried out in artificial conditions show that this property is fairly general and that an animal may be conditioned to periods which bear no relationship to natural rhythms, provided that these cover less than 24 hours. Proof of this has been furnished mainly by the work of Pavlov, his pupils, and his school, and is to be found in Pavlov's book *Conditioned reflexes* and in a recent article by Dmitriev and Kochigina (1959). As early as 1907, Zelenyi subjected a dog to a combination of sound and food every ten minutes and found that after a certain number of repetitions, the conditioned salivary reflex occurred regularly every ten minutes. In 1912 Feokritoff investigated the precise laws governing this phenomenon. We shall return to this later (pp. 29 ff.). But this phenomenon is not limited to the salivary reflex. Beritov (1912, quoted by Dmitriev and Kochigina, 1959) showed that it is also possible to condition defense motor reactions. After an electric shock has been applied about 40 times at 5-minute intervals to the front paw of a dog, it will be seen to stir a minute before the next stimulus is due, move its head and lift its paw. This reflex is not established all at once, however. At first the reactions are spread out over the whole interval between two stimuli and they only gradually begin to concentrate toward the end of this time. Finally, after 1936, Bykov and his colleagues (quoted by Dmitriev and Kochigina, 1959) proved that it is possible to condition the changes of metabolism to time (e.g., reactions to changes in temperature). These reactions are all comparable to those of Beling's bees.

Periodic changes induce a rhythm of behavior. It is even possible to train animals to follow a complex periodicity. If pigeons are fed according to the following timetable: feed, 15 seconds; pause, 30 seconds; feed, 15 seconds; pause, 90 seconds; and this cycle is repeated several times a day for several consecutive days, it is found—by actographic recording—that they remain quiet during these pauses but begin to move toward the end of them, in anticipation of the arrival of food. If feeding is discontinued, they will continue to behave in the same way for several more cycles. (Popov, *Études de psychophysiologie*, 1950.)

Man has been relatively less studied than animals or plants from the point of view of adaptation to cosmic periodicity. To begin with it is more difficult to experiment on humans and to subject them to totally artificial conditions. Moreover, as will often be stressed, their methods of adaptation are varied; they can become strengthened, or also compensated, to the point of obscuring basic facts.

Casual observation is enough to show us the importance of the diurnal rhythm in our lives. Nearly all humanity sleeps at night and works during the day. This is another case where the rhythm of light shapes an organic necessity, for no one can do without sleep after activity, even though the periodicity of the alternation can be varied to a considerable extent in some exceptional cases. Within this general framework, adaptations remain individual. Many people wake at a more or less fixed time and any accidental and appreciable variation in their bedtime makes no difference in this. In this case the time of waking is not determined by the quantity of sleep but by habit. Since air travel has made it possible to go quickly from one country to another separated by several time zones, many people have noticed that they have difficulty in sleeping for several nights after such a journey; coming from France for instance, they tend to wake up much too early at the beginning of a stay in America (where the sun rises five hours later).

It has been known for a long time that the pulse, blood pressure, and especially temperature of the body present day-night variations in humans as well as in many animals. There is about 1.8°F difference in the human temperature between the minimum at night and the maximum in the afternoon. In 1875, physiologists already attributed this rhythm to the alternation of light and darkness which brought with it the alternation of activity and rest; they therefore thought it possible to reverse this by substituting nocturnal for diurnal activity. The results of their experiments remained very controversial, however, until Toulouse and Piéron found in 1907 that the temperature change was reversed in the case of nurses changing from day to night duty. This reversal was gradual and was not completed until after 30 or 40 days. During the first few weeks the rise in temperature, which usually takes place in the mornings and the early part of the afternoon, grew gradually less marked, until it finally changed to an increasingly rapid drop (Toulouse and Piéron, 1907).

Travellers on long sea journeys have noticed a gradual change in the rhythm of their temperature. For instance, Osborne, having left Melbourne by ship for England, noticed after six weeks at sea that his temperature still reached its maximum every day at about 6:00 P.M.; on his arrival this corresponded to 4:00 A.M. in Melbourne. The reversal was complete. (Osborne, 1907.)

"The fact of reversal proves that the rhythm depends on the conditions of life, on physical and mental activity which normally reaches a peak at a moment determined by cosmic conditions, that is by the light of the sun; there are, however, modifications of a social nature which explain the fact that this is reached much later in towns than in the country. It is therefore impossible to speak of one fundamental periodicity.

"On the other hand the difficulty and slowness of reversal show that the rhythm has been firmly established and tends to maintain its periodicity and oppose the introduction of a new periodicity. Thus a compromise arises gradually between the past,

remembered action, which grows weaker, and the present action, which increases in strength." (Piéron, *L'évolution de la mémoire,* 1910, pp. 89–90.)[4]

The controversy which preceded the experiments of Toulouse and Piéron did not cease immediately, for some of their results were contradictory. It seems that this reversal does not occur in certain individuals (Regelsberger, 1940), whereas it is always possible in the case of animals. These exceptions may perhaps be explained by some internal factor of a psychic nature. It does indeed seem that when the psychic life weakens, malleability by external influences increases. In the case of a microcephalo-acromegalic oligophrene, a double daily reversal of temperature was obtained in only six days by exposing him to light from 6:00 to 12:00 A.M. and 6:00 to 12:00 P.M. and keeping him in the dark from 12:00 to 6:00 A.M. and P.M. His temperature dropped between midnight and 6:00 A.M., rose from 6:00 A.M. to midday, dropped again from midday to 6:00 P.M. and rose again from 6:00 P.M. to midnight. However, during the month of the experiment there were appreciable irregularities and even a recurrence of simple reversal, i.e., of the more fundamental rhythm of 12–12 (Burckard and Kayser, 1947).

It is not only possible to reverse the variation in temperature, but also in certain cases to achieve a new rhythm of the temperature curve at periods slightly different from the circadian rhythm. Kleitman made one of his helpers live an 8-day week (15 hours awake, 6 hours asleep), then a 7-day week (17 hours awake and 7 asleep), then a 6-day week (19 hours awake and 9 hours asleep). In all three cases the temperature curve followed the

---

[4] These laws have been confirmed by tests carried out on night workers. Those who change their times of work every week have an irregular temperature. Women who always work at night, such as some night nurses, sleep badly the first few nights of their annual leave. This shows that they need a certain amount of time to readjust themselves to another rhythm of activity and rest.

As an indirect consequence of adaptation to periodic change, night workers whose physiological rhythms are always being upset by changes in their conditions of work have been found to be most prone to diseases connected with autonomic disturbances, e.g., neurosis, anxiety states, respiratory and digestive disturbances (quoted by Kleitman, 1939, and Neulat, 1950).

rhythm of activity. It must be added, however, that when Kleitman repeated the experiment on himself, it failed (Kleitman, *Sleep and wakefulness as alternating phases in the cycle of existence*, 1939).

Apart from the rhythm of temperature, the rhythms of a number of physiological functions are also connected with phases of activity: the blood sugar level, the calcaemia and proteinaemia level, the lymphocyte count, renal secretion, the biliary and glycogenic functions of the liver (Kayser, 1952). Thus the entire human organism, both conscious and subconscious, is affected by the day-night rhythm.

These rhythms are acquired. Nothing proves this better than a study of the development of a child. No form of circadian cycle is to be observed in the foetus or the new-born child, either in its activities or in its physiological functions. The sleep of a new-born child is polyphasic; the periods of sleep are very numerous and there is no preference for day or night. As he develops, the periods of sleep become longer and less numerous. It is, however, the social habits connected with the succession of day and night which play a major part in the timing of these periods of sleep and wakefulness; at the end of one week of life, nocturnal sleep is already more important than diurnal (Gesell, *The embryology of behavior*, 1943). The temperature rhythm is established more slowly; it only becomes well defined in the course of the second year (Kleitman, Titelbaum, and Hoffmann, 1937). The main point to be noted in connection with these inductions seems to be the establishment of a periodic rhythm of activity and not the direct influence of an agent such as light or darkness. It should be noted that exactly the same rhythms are followed by congenitally blind people (Remler, 1949).

Adaptation to periodicity is also apparent in the cycle of meals. In Europe, if not elsewhere, babies are accustomed from birth to 3-hourly feedings with one feeding omitted in the night. Most healthy children adapt very quickly to this complex rhythm, the process being complete after a month. Marquis (1949) has made some experiments concerning this adaptation and has shown some very interesting facts. Three groups of babies were placed at birth in actographic beds through which their activity was recorded. The first group was fed according to the babies' natural rhythm, i.e., every time they cried of hunger. This natural rhythm averaged 3 hours 2 minutes. The second group of babies was fed regularly every 3 hours, and the third every 4 hours. In the latter two groups, the curve of activity of the babies was regular by the end of the first week. Activity decreases after the feeding, reaches a minimum and begins to increase again before the time of the next feeding. The activity of the babies accustomed to a 4-hour rhythm increases more than that of the groups with a 3-hour rhythm; this is normal, for the natural rhythm of the need for food appears to be slightly more than 3 hours but certainly less than 4. In the case of the children accustomed to a 4-hour rhythm, activity gradually began later; by the sixth day it was starting after 3½ hours. On the ninth day the second group was changed from a 3- to a 4-hour rhythm. They became very active between the third and fourth hours of every cycle, far more so than the babies who had been fed every 4 hours from the beginning. This phenomenon shows very clearly that they had already adapted to the 3-hour rhythm. This very swift adaptation to the 3-hour and the 4-hour rhythm raises the question, to which we shall later return, of the underlying physiological mechanisms. We know that integrations at the cortical level are not possible in a new-born child because the fibers of the cortical cells are not myelinated and the network of their interconnections is not yet developed. Marquis thinks that a subcortical center regulates this very primitive form of adaptation.

The periodicities we have so far considered are connected with the main biological activities of the organism, namely motility, rest, and food. In every case an organically necessary alternation synchronizes with periodic changes in the outside world. It is also possible, however, in the case of humans, animals, and plants, to induce rhythms which bear no relationship to organic alternations. If a person receives a slight electric shock, a reflex reaction is seen, known as a psychogalvanic reaction; this is caused by a reduction in the apparent resistance of the skin which is itself brought about by a sympathetic activation. If this shock is repeated every 8 seconds for a certain length of time, when the shocks are ended one or more reactions will occur in certain individuals at intervals of about 8 seconds. This proves that a rhythm of autonomic reactions has been induced (Fraisse, and Jampolsky, 1952).[5]

All the typical examples we have chosen show quite clearly that induction of periodic changes on the level of physiological reactions or of activity, is a general law for all organisms. The rhythms of the natural environment give rise to rhythms in the organism which are at first exogenous but which actually become endogenous; its behavior anticipates the presence of the stimulus and the rhythm continues for some time after its cause has been removed.

To gain a fuller understanding of this phenomenon, we must now consider its underlying mechanism.

II *The Laws of Periodic Adaptation*

To understand the induction of rhythms by periodic changes, we must remember that rhythmic oscillation seems to be a characteristic of the functioning of the nervous system. No doubt this property extends to other tissues, especially in the less complex organisms. Our aim is, however, to understand human behavior and we can therefore pass over this aspect of the phenomenon. There are three rapid endogenous rhythms which are particularly important: those of the heart, of breathing,[6] and of the electric activity of the brain. In all three cases it has been proved that these rhythms are not periodic responses to periodic stimuli; the only effect of the stimuli which act on them is to accelerate or slow down the spontaneous oscillation of the nervous centers.

It has also been found that nervous tissues which are not

---

[5] In actual fact this phenomenon is more complicated. Repeated electric shocks give rise to a double series of psychogalvanic reactions: reflex reactions to the shock itself and reactions which precede the shock and which we have called anxiety reactions. When the shocks have ceased it is found that induction has taken place with both kinds of reaction. Reflex reactions are better defined in cases where less anxiety reaction occurs.

[6] In fact it appears that the nervous centers of respiration have their own rhythm, which is only "controlled" by the carbonic content of the blood. Adrian and Buytendijk (1931), for instance, observed alternations of activity in the respiratory centers of fish even when these centers were shielded from any influence from the varying oxidation of the blood.

spontaneously rhythmic do, however, respond rhythmically when they are exposed to a constant stimulus. This is the case for the reflex centers, the sensory fibers, and motor fibers (Fessard, 1931, 1936). In all cases both centers and fibers show that they have their own period of response and the rhythm of response only corresponds within certain limits to the rhythm or simply to the intensity of the stimulus.

A good example of this is the scratch reflex of the dog. As Sherrington has shown, a simple stimulus can set off a series of periodic movements which cannot be explained by successive stimulation, as the reflex still occurs even if the afferent pathways of the muscles concerned have been severed. We must conclude, therefore, that the rhythm of the movement, whose frequency is independent of the nature of the stimulus, is explained by the repetitive activity of a center (*The integrative action of the nervous system*, 1906, pp. 45, 71-122).

One of the most interesting points for the comprehension of the phenomena of adaptation is the tendency of these nervous rhythms to synchronize with each other. The periodicity of one part very often acts as a pacemaker for other pulsations. Fessard (1936) has shown the existence of this phenomenon in the nervous conducting paths. We know that the sinus node in the heart is considered to be a pacemaker for a number of other centers which all have their own periodicity. In the higher centers, the periodic cerebral waves recorded by electroencephalography are the result of a widespread synchronization of the electrical activity of the nerve cells. According to the most acceptable hypothesis, the regularity of the pulsations of any organ or center is mainly due to the coordination of a large number of elementary pulsations (Bethe, 1940). Even more important is the fact that certain periodic activities can become synchronized with stimuli which are themselves periodic. Thanks to the research carried out by Adrian (1934), whose results have since been confirmed many times, we know that the alpha rhythm of cerebral waves can be regulated to a certain extent by an intermittent light.

Synchronization is also frequently apparent in alternating movements. These owe their regularity to the phenomenon of successive induction. The contraction of the flexors leads to the contraction of the extensors, and so forth. This succession has its own tempo, as has been proved by a number of experiments concerning the spontaneous tempo of mastication, of walking, of the swinging of a limb or the swaying of the trunk, etc. The most interesting fact established is that these alternating movements can be induced by rhythmic stimuli. This induction can be effected in a child of nine months, and it is this same aptitude which makes it possible for large military formations to march in step, although the individual tempo may vary from one soldier to another.

These facts all concern relatively rapid rhythms, but through them we can understand by analogy those cases of induced behavior which have a longer period. They demonstrate two properties of the nervous system. First, the tissues and especially the nervous centers either have a spontaneous rhythmic activity or they show a natural rhythmic response to stimuli; this explains the fact that inductions are frequent and easy to establish. Second, whether spontaneous or triggered, the rhythmic activity of a center has its own particular frequency which is only modified within certain limits by any regulation, stimulus, or synchronization that may occur. We have seen in particular that diurnal rhythms, although induced by the alternation of day and night, correspond to an optimum frequency for organisms, for it is difficult to establish rhythms of this type with longer periods.

The analogy which can be drawn between rapid and slow rhythms and between endogenous and induced rhythms is justified by the fact, verified by research, that variations in temperature have the same sort of effect on the first as on the second. The effect of temperature is to increase the speed of chemical reactions. The exponential law relating the speed of reaction to the absolute temperature was found by Van 't Hoff in 1884; following this, Arrhenius was the first to show that the same law applies to biological processes (Sivadjian, *Le temps,* 1938, p. 349). The logarithm of the frequency of reaction is inversely proportionate to the absolute temperature, according to the formula:

$$\log f = c - (\mu/2.3 \, R \, T)$$

(*f* is the frequency or speed of reaction, *R* is the constant for perfect gases, and *T* the absolute temperature).

For every type of reaction there is a constant μ called the *temperature characteristic* or *thermal increase,* which characterizes the activating energy of the process in calories per molecule-gram (Hoagland, 1936*d*).

Since the work of Marey (Sivadjian, *op. cit.,* p. 346), it has been known that the refractory period of centers and nerves depends on the temperature, but it has been proved that the endogenous rhythms of the organism also follow Arrhenius' law. The heart rate is accelerated by an increase in the temperature of the body. The constant μ then has a value of approximately 29,000 in humans. Respiration is modified in the same way, the value of μ varying with the species. In the same way the alpha wave rhythm is accelerated under these same conditions, the value of μ being 8000 in normal cases and higher in cases of general paresis (Hoagland, 1936*a, b, c, d*). The frequency of the rhythmic discharges of the nerves is also governed by the same law; Fessard calculated a constant of $\mu = 14,900$ for the nerves of crustaceans (*Recherches sur l'activité rythmique des nerfs isolés,* 1936, p. 135).

The period of circadian rhythms is also found to decrease when the temperature is raised. This can be seen in the periodic movements of the leaves of the bean plant (Bünning, quoted by Piéron, 1937). The same is found to be true for induced rhythms. Bees trained to come for food at a certain time will come early if the temperature is raised and late if it is lowered (Wahl, 1932). It is true that Stein (1951) found that pigeons trained under similar conditions were not affected by changes in the temperature of their surroundings, but the pigeon is a warm-blooded animal and changes in the temperature of its environment have no appreciable effect on its body temperature, which is all that matters in this case.[7]

[7] However, Marx and Kayser (1949) found that the circadian rhythm of the activity of the lizard, a cold-blooded animal, was not modified by variations in temperature if the animal was kept in complete darkness. The period of activity was simply three times longer at 29° than at 19° C, and the center of the phase of activity remained unchanged. This case does not, however, invalidate the facts given above because there must be some

These results not only show that innate and induced periodic regulations have the same characteristics, but they also explain the problems of what may be called the estimation of duration.

Well before any of this research, Piéron (1923) had already come to the conclusion that our appreciation of duration might be dependent on physiological processes. ". . . And if the speed of organic processes is modified, by variations in temperature for instance, mental time will increase or decrease proportionately." This hypothesis was to inspire the well-known experiments of François (1927, 1928). A subject is asked to strike a key at a subjective rate of three taps to the second. His body temperature is then raised by diathermy. An acceleration in the tapped rhythm is observed, of which the subject is obviously not aware. The same result is obtained if the subject is asked to count at a rate of one number per second. Hoagland (1933) found the same results in the case of persons whose temperature was increased by illness; according to his calculations these results and those of François confirm Arrhenius' law, μ having a value of 24,000.

The experiments carried out by François and Hoagland may seem entirely different from those previously described where an acceleration in biological rhythms was found when the temperature was raised. The former do in fact concern voluntary rhythms of movement, but to take the duration of one second as a reference requires not an exact knowledge but a norm determined by the individual experience of each subject. The timing of a recurring movement is regulated at will, but the norm to which it refers is established by processes which cannot be governed in the same way. Like the bee which arrives early when it is warmer, the human whose temperature is raised will tap faster without realizing that he has changed his tempo.

The fact that all rhythms, both innate and induced, are governed by the same law[8] shows that the mechanisms of their

---

other kind of special mechanism of regulation in the lizard, probably humoral; Hoffmann (1955) found, as mentioned above, that the circadian rhythm of lizards is innate.

[8] The pharmacodynamic effect of certain substances must be considered in addition to the influence of temperature. Grabensberger (1934) showed that bees and wasps trained to a 24-hour rhythm arrive early on the follow-

regulation must be similar. Nevertheless, the differences in the value of the thermal increase (μ) make us think that no endogenous oscillation, whether of the heart beat, of breathing, or of alpha waves, can act as a cue or, more precisely, as a pacemaker for adaptations to periodic changes. These have their own regulating mechanism. What is it?

At present we can do nothing but speculate on this point. We still cannot even explain the innate rhythms themselves. Why do certain centers have a spontaneous periodic activity and why do the nerve fibers respond periodically to stimuli? As Fessard has pointed out, we must make a distinction between the process of *activation*, which makes a neuron capable of periodicity, and the process of *excitation*, which merely releases it. The two are favored by different conditions. We cannot explain the property of automaticity simply by the existence in the nervous centers and nerve fibers of the refractory phase which follows any period of activity. In fact, the duration of the refractory phase is not proportionate to the actual period of the rhythm, the latter being far longer than is necessary for simple physiological reconstitution. Fessard therefore assumed the existence of a phenomenon of *self-excitation* (*Recherches sur l'activité rythmique des nerfs isolés*, 1936, pp. 144-145).

As an explanation of cyclic inductions, Pavlov suggested the successive mechanisms of excitation and inhibition, controlled by conditioning processes. Although he admits that this explanation is still too vague, he does not believe that there is any more to induced rhythms than to classical conditionings. Time, he says, acts as a conditioned stimulus, but obviously the question is not that simple. Disregarding the cyclic phenomena of the external world, time manifests itself by a series of periodic organic changes. Pavlov draws the following conclusion from their existence: "If we take it that every state of the organ in question can

---

ing day if they are dosed with iodothyroglobulin, which accelerates cellular exchanges. On the other hand they arrive late if dosed with quinine. Sterzinger (1935) found that the spontaneous tempo was slowed down in the case of subjects who had taken quinine and accelerated when they had taken thyroxine. This points to yet another link between induced rhythms (in animals) and spontaneous rhythms (in humans).

affect the cerebral hemispheres, this may act as a basis for the distinction of one moment from another." In other words, if the organs send messages of a different nature according to the moment, certain of these will become conditioned stimuli, if they are associated with an unconditioned stimulus (such as food), i.e., their return will create a stimulus, while others, which are not reinforced, will become conditioned inhibitors.

In Feokritoff's experiment (see p. 36), where the dog is fed every half hour, this feeding "was accompanied and followed by a definite activity in a large number of organs, which all underwent a series of definite cyclic changes. All these changes were reflected in the cerebral hemispheres where they fell on appropriate receptive fields, and at a definite phase of these changes acquired the properties of the conditioned stimulus." (*Conditioned reflexes*, p. 43.)

There are two aspects to Pavlov's interpretation. First, according to him, phases of excitation and inhibition alternate in the center concerned during an adaptation to periodic changes. Second, these successive phases are regulated by the periodic activities of organs whose messages take on a different significance at the hemisphere level according to the stimuli with which they were associated during the training itself. This explains the first spontaneous salivation of a dog when it is given nothing to eat after being fed every 30 minutes, but it is an inadequate explanation for all the cases where rhythmic induction continues for several cycles in the absence of any reinforcement. Alternatively we can assume what should be explained: that periodic stimuli cause periodic changes which may be repeated for several cycles without any reinforcement. Thus the body temperature or any other neurovegetative manifestation may become a source of conditioned stimuli; first, however, we must establish the mechanism through which these organic rhythms have themselves been induced and the way in which they are maintained. Nevertheless, it is true that from the moment of coming into existence, these mechanisms can themselves act as a basis for other forms of temporal conditioning, which will then be of a secondary nature.

There is a solid experimental basis for this description in terms of excitation and inhibition. Let us review the facts. Feokritoff (1912), in his research on time as a conditioned stimulus, observed the same phenomena as described by Pavlov: "The animal can be given food regularly every 30th minute, but with the addition, say, of the sound of a metronome a few seconds before the food. The animal is thus stimulated at regular intervals of 30 minutes by a combination of two stimuli, one of which is the time factor and the other the beats of the metronome. In this manner a conditioned reflex is established to a compound stimulus consisting of the sound plus the condition of the hemispheres at the 30th minute, when both are reinforced by food. Further, if the sound is now applied not at the 30th minute after the preceding feeding, but, say, at the 5th or 8th minute, it entirely fails to produce any alimentary conditioned reflex. If it is applied slightly later it produces some effect; applied at the 20th minute the effect is greater; at the 25th minute greater still. At the 30th minute the reaction is of course complete. If the sound is never combined with food except when applied at the full interval, in time it ceases to have any effect even at the 29th minute and will only produce a reaction at the 30th minute—but then a full reaction." (Pavlov, op. cit., p. 42.) The process of inhibition is obvious; the additional stimulus only comes into effect when the cortex begins another period of excitation under the influence of the time factor. The work of Koupalov (1933), another of Pavlov's pupils, is even more revealing. He created a double conditioned reflex to mechanical excitation of the skin. At one point the reflex is positive (salivation) and at another negative. A stimulus is applied to these two points alternately every 7 minutes. If a negative stimulus is given 14 minutes, instead of 7, after a positive stimulus, it gives rise to as much salivation as the latter. Inversely, if a positive stimulus is applied 14 minutes after a negative one, it has less effect than it would have had after 7. On the other hand if applied after 21 minutes it has the same effect as after 7. These results are easy to explain if we assume that phases of excitation and inhibition are alternating in the brain every 7 minutes. Whether the positive stimulus is applied 7 or 21 minutes after the negative stimulus, its effect is positive and of the same strength, because it coincides with a phase of cortical excitation. If a negative stimulus is applied during this same phase, it has some effect (about half). Similarly, positive stimuli applied during phases of cortical inhibition have a very reduced effect. The Russian school has also shown that injections of bromide, which facilitate processes of inhibition, also promote the establishment of time reactions (Deriabin, 1916; Bolotina, 1953; Kochigina, quoted by Dmitriev and Kochigina, 1959). Furthermore, the establishment of temporal conditioning and its resistance to external disturbances depend on the type of nervous system of the dog. Better results are obtained with animals in which inhibition processes predominate than with excitable animals.

Laws of the same kind have also been established by some American authors, who base their reasoning on more operational notions. If rats are given an electric shock every 12 seconds, they will jump with measurable force at every shock. When this training has been carried on for some time, they are given a shock 3, 6, 9, 12, 15, 18, 21, or 24 seconds after the previous one, (the timing being selected at random). It will be found that the rat jumps with most force at an interval of 12 seconds and with decreasing force as the interval grows larger or smaller. Brown (1939), who conducted the experiment, explains this by Hull's law of the goal gradient or reinforcement gradient (1932). The nearer a stimulus comes to the correct point (in time or space), the stronger the reactions it causes.

To put this in more physiological terms, the stimulus is weaker (or the inhibition stronger) when the shock is given before or after it is "expected." The same conclusion can be drawn from the results obtained by Rosenbaum (1951). He used instrumental conditioning (see Chapter 2, pp. 54 ff.). The rats were trained to push a lever as soon as it was shown to them in a Skinner box. They were shown the lever every 60 seconds at first, then at longer or shorter intervals when they had been trained. The force of the reaction was measured by the amount of time which elapsed between the animal's being shown the lever and its pushing it. This latent period is negligible (about a second) when the lever appears 60 seconds after the previous reaction. It increases progressively when the interval is longer or shorter and reaches about 5 seconds when the interval is only 15 seconds.

These facts all point in the same direction. The duration is registered in such a way that the reaction is strongest at the moment when it is "expected," in view of the preceding training. When the interval is longer or shorter, the reaction gradually becomes weaker, probably as the result of a process of inhibition.

This brings us back to the main problem: how is the alternation of periods of excitation and inhibition regulated temporally? Following in Pavlov's footsteps, Koupalov, like Frolov (1935), inclined to explain this by external or internal stimuli associated with the conditions of work. We have already shown that this explanation is doubtful, and we would add that it is not necessary on theoretical grounds, because the nervous system is known to have an intrinsic rhythm, even when responding to stimuli that are not periodic. Popov, who made a thorough study of these problems, thinks that we must assume a specific property of the nervous system "of reproducing activations in the same order as that in which they were originally aroused by the corresponding stimuli." He proposes giving this property the name *cyclochronism* (*Études de psychophysiologie*, 1950, p 17). This was suggested to him by observations similar to those we have described, for he found in his experiments that the responses were not only periodic but in some cases actually stereotyped. He observed this in the activity of pigeons fed at regular intervals (see p. 24); he also found it in the electroencephalographic recording of a rabbit which had a light shone periodically in its eye: the modifications in the visual cortex become periodic and continue more or less the same after the stimulus has ceased (*ibid.*, p. 15). According to Popov, who agrees with Fessard on this point, we must admit that the response of the nervous system to a stimulus is not simple but polyphasic. A phase of inhibition follows on a phase of excitation and may be followed by other more or less complex phases of excitation and inhibition (*ibid.*, pp. 18, 62–63).

These induced rhythms are then the result of a conditioning process in which the interval of time between two periodic stimuli is the conditioned stimulus, so that when one of the periodic stimuli is omitted a reaction still takes place. As in the case of conditioning, this acquired organization requires repetition; if there is no reinforcement it will gradually disappear.

Even if cyclochronism does seem to be a basic property of the nervous centers, we still need to find out whether it applies to all or only some of the centers. This question also remains unanswered and can only be tackled indirectly. If it is true, as Pavlov thought, that all conditioning, at least as regards the higher vertebrates, takes place at the cortical level, it still seems that periodicity is also established in the subcortical centers. Deriabin has in any case shown that temporal conditioning remains after removal of the region of the sensory cortex concerned with the stimulus associated with food (tactile or auditory area).

It seems likely, nevertheless, that the cortex is necessary for the establishment of conditioning, although this has not been proved for temporal conditioning. On the other hand certain observations show that the periodicity of the induced rhythm can be controlled by subcortical centers. Kayser (1952) showed that pigeons whose cerebral hemispheres had been removed still followed the day-night temperature rhythm, which is acquired. The regulation of sleep depends on the activity of a hypothalamic center situated near the infundibulum and the base of the third ventricle; a decerebrate dog still shows the normal alternation of waking and sleeping (Lebedinskaia and Rosenthal, quoted by Fulton, *Physiology of the nervous system*, 1947, p. 547). It may also be thought that the rhythms of motor as well as purely vegetative activities are governed by the basal nuclei, since these control automatic movements.

It is generally agreed today that the hypothalamic region is responsible for the regulation of organic cycles; from this Kleist (1934) and subsequently Klines and Meszaros (1942–43) have drawn the conclusion that the temporal integrations of periodic reactions also take place at this level. We shall see in Chapter 6, p. 163 the facts concerning temporal disorientation in Korsakov's syndrome which bear out this hypothesis.

III *Orientation in Time*

Rhythmic induction, or the occurrence of organic periodicities which synchronize with periodicities in nature, constitutes a form of adaptation to the temporal conditions of existence. The general biological significance is obvious. Rhythmic induction permits living creatures to turn reflex reactions into reactions of anticipation. Thus the *Convoluta* can burrow down into the sand before they are covered by the rising tide, while sea anemones close before the tide ebbs, thus keeping their water and avoiding the dessication which could be fatal to them (Piéron, *L'évolution de la mémoire*, 1910, p. 74). The bee which has discovered a source of nectar can find it again more easily the following day and also adapt itself to the time of secretion of this nectar, as this occurs at fixed times, differing from flower to flower. If bees are given water containing different concentrations of sugar at different times of the day, after a few days most of them will arrive at the time when the strongest concentration is due (Wahl, 1933).

In fact, the necessary adaptation of the organism to the periodic changes it undergoes, and in particular to the alternation of night and day, is effected more economically thanks to this internal regulation. Proof of this is given by the fatigue occasioned by the readaptation of the organism to a new rhythm of activity. We have already seen that people flying from Paris to New York or vice versa had difficulty in adapting to the new hours of sleep for several nights, owing to the persistence of the old rhythm.[9] Complete inversion of the rhythm of activity causes such fatigue that the first authors who tested on themselves its influence on the diurnal temperature rhythm abandoned their efforts after a few days (Piéron, *ibid.*, p. 88). Doctors and sociologists have also often noted the difficulty which many workmen have in changing their rhythm of work, for instance

---

[9] We should compare these observations with the behavior of an orang-outang. In Java, where it was captured, it slept regularly from 6:00 P.M. until 6:00 A.M. When taken by boat to Germany, it continued to sleep and wake as if in Java; thus at the longitude of the Cape of Good Hope it was sleeping from 2:00 P.M. to 2:00 A.M. (Groos, 1896).

when they change from a day to a night shift. (Neulat, 1950.)

The existence of organic rhythms induced by periodic variations in the environment has particularly important psychological consequences for man. They provide him with an internal clock.

What is the use of a clock? First, to bring every minute of the day into relationship with the whole course of day and night, or more scientifically, to determine the relative position of the earth and the sun during the course of the day-night cycle. Nature acts as a clock when we are content to tell the time, for example, from the position of the sun or the shadows it casts (sun dial). Man himself has also constructed various kinds of clock to show periodic movement. On some of these the hand moves once round the dial every 24 hours. From the position of the hand, as from that of the sun, we can tell the time, or the division of the day. Man uses these means systematically to organize his work and his leisure, and in particular to coordinate his activity with that of his fellows.

Man has, however, a certain sense of time independent of these objective cues, which is most obvious when none of the latter are available. He acts then as if he were capable of interpreting organic messages whose significance, being associated with the periodic modifications of the body, is related to the time on the clock. W. James quotes the case of an idiot who could not tell the time but asked for her soup every day at exactly the same hour (*Principles of psychology*, 1891, p. 623). The mentally ill who are "temporally disoriented" are actually only disoriented as regards the conventional time of calendars which divide the year into months and number the years from an arbitrary point in time. We have confirmed that they are not disoriented as to the hour of the day; it is true that hours do represent an arbitrary division of the sidereal day, but they are reinforced, as symbols, by the regularity of hospital life: the hours of bed making, visits, meals, lights out, etc. Between these events the patient can still be guided by the fundamental organic rhythms of food and sleep, and more generally of metabolism. In every case, except where complete dementia makes the question meaningless, patients considered to be temporally disoriented are capable of giving the time to within 60 minutes, and this degree of accuracy is no lower than that of many normal adults (Fraisse, 1952*b*). These examples of behavior are all taken from a hospital and are obviously open to discussion. But several experiments, which we are about to examine, prove man's ability to orient himself in time, to a certain extent, with no help from external cues; that is, to make use of an internal clock. These are all connected with an estimation of the time on waking, when the number of external cues available is at a minimum.

1. The Experiment of MacLeod and Roff (1936)

In order to examine the accuracy of our temporal orientation without help from the cues provided by nature or civilization, these two authors took turns in shutting themselves up in a soundproof, air-conditioned room lighted only by artificial light. They had food, a bed, and a lavatory constantly at their disposal. They were to do nothing except perhaps make notes on their observations. From time to time—except while sleeping—they were to pick up the telephone and say what time they thought it was. The first man stayed in the room for 86 hours, that is nearly 4 days. At the end of this time his estimation was only 40 minutes out, but he had previously erred by 4 or 5 hours. The second subject remained in the room for 48 hours, at the end of which he was 26 minutes out, his maximum error having been 2 hours. To what can we ascribe this relatively great accuracy? If we study the experimental procedure of these two authors— particularly of the first subject, who stayed longer in the room— we find that the basis of their temporal orientation (and of their temporary disorientation) was their estimation of the time on waking. The first subject began the experiment in the evening and went to bed at about midnight, but he woke at 4:43 in the morning, believing it to be 9:00 A.M. He slept badly, as was to be expected in this unusual situation. That was his greatest error. He compensated by having a sleep in the afternoon, but still went to bed early and woke up at 10:50 A.M. He estimated then that it was 10:00 A.M. After the third night he woke at 11:28 A.M. and estimated that it was 9:00 A.M. Two facts should be noted: (a) Despite the artificial living conditions, the subject went to bed and got up at more or less normal hours; (b) When he slept normally his time judgment on waking was not far out. After that he was sure of his orientation during the day, this being facilitated by the rhythms of meals and bodily needs.

It is therefore evident that the rhythm of sleep acted as a clock for these subjects. We know that as our bedtime draws near every evening we feel sensations of tiredness which act as cues, and in the morning we wake at a more or less fixed hour.[10] This experiment shows, however, that our judgment may vary with the time of waking. To show this, other more specific experiments have been carried out. We shall examine these.

2. Time Judgment After Sleeping

Our estimation of the time when we wake up in the morning under normal conditions and after a good night's sleep is not a good example of the influence of internal cues because we then have a number of outward signs at our disposal, such as the light of day and the noises from our surroundings, which are in themselves as good as a clock. However, when an unusual noise or a nightmare wakes us up in the middle of the night, we immediately have some idea of what time it is. To give more precise observations, L. D. Boring and E. G. Boring (1917) woke volunteer subjects systematically between midnight and 5:00 A.M. and asked them what time it was. The average error in judgment was 50 minutes; this clearly shows temporal orientation. The subjects then try to interpret their own internal sensations: degree of fatigue, depth of sleep at the time of waking, sensations in the stomach, degree of fullness of the bladder. They may also be interpreting unconscious cues, the kind which make us say, without calculation, when suddenly wakened, "It must be three o'clock."

3. Spontaneous Waking at a Specified Time

Many people insist that they are able to wake up at a time they have decided upon the night before. Is this belief without

---

[10] The same has been found to be true of birds. They wake at their customary time even if kept away from any noise or light. Canaries, for instance, wake at a fixed hour whatever the conditions (Szymanski, 1916).

foundation? Or if the ability really does exist, how can we explain it?

Serious research has confirmed the existence of this ability and explains it by the action of the physiological clock. It is obvious straight away that not everyone possesses this ability. Clauser (*Die kopfuhr*, 1954) has made the most thorough investigation of this question. Of the 1080 people he questioned, 19 percent said they were unable to wake up at will, 29 percent were doubtful and 52 percent had occasionally observed this phenomenon in themselves. Of this 52 percent, only 15 percent could always rely on their ability, 20 percent usually could, and 59 percent could only from time to time. On the whole, the ability seems to be fairly evenly distributed, with two extreme groups, 18 percent being more or less sure that they can wake up and 19 percent being absolutely incapable of doing so. This variation from individual to individual is confirmed by the research of Omwake and Loranz (1933) carried out on two groups, one consisting of 10 students who claimed to be capable of waking up at any hour and the other of 10 who said they were not. For 14 nights these girls were asked to wake themselves up at different times, varying from 12:30 A.M. to 6:15 A.M.; in the first group 49 percent of the cases were successful to within 30 minutes. In the second group only 5 percent were successful. There are therefore marked differences from one individual to another. On the whole there is more information available on the subjects with the greatest aptitude for waking up automatically at a fixed time. Hall (1927) carried out 109 tests on himself; in 18 percent of these cases he woke up at exactly the right time, in 53 percent within 15 minutes of the time, in 75 percent within 30 minutes and in 81 percent within 54 minutes. Brush (1933) attained approximately the same degree of precision. In 50 tests carried out on himself, his mean error was 10.6 minutes with a σ of 10 minutes. Vaschide (*Le sommeil et les rêves*, 1911) found a mean error of 21 minutes for 33 subjects and Frobenius (1927) found an error of less than 40 minutes in 96 percent of the tests he carried out on 5 subjects over 250 nights.

These results are very similar and show considerable precision. A number of experiments have shown, however, that it becomes easier to wake up at a fixed time and to be more accurate the nearer this time is to the normal time of spontaneous waking. Vaschide chose 17 subjects who were in the habit of waking up at 8:00 A.M. and carried out 19 tests on each one. He found that their errors averaged 28 minutes when they woke at 3:00 A.M., 23 minutes at 5:00 A.M. and only 17 minutes at 6:00 A.M. He also experimented on himself for 257 nights and found that his mean error was 25 minutes when waking at 1:00 A.M. and only 6 minutes 50 seconds when waking at 8:00 A.M. Similarly, Omwake and Loranz had 36 percent successful results (with a maximum error of 30 minutes) when subjects who claimed to have this ability were tested for waking at times before 2:30 A.M. and 70 percent at times after 4:30 A.M.

These facts being definitely established, how can we explain them? The first possibility which comes to mind is that these people make use of external cues: a cock's crow, the noise of a train or a factory siren, the first bus, etc. Vaschide's subjects agreed that they tried to use these but said that they were inadequate. Frobenius, who made a special study of this problem, found that the results from his subjects were equally accurate whether they had a clock striking the hours in their bedroom or slept in a room very isolated from noise. He also found that the degree of accuracy with which subjects woke up was not lessened if there was a clock in the room which was set wrongly without their knowledge.

The way in which subjects wake up under these conditions shows that it is not always possible to use external cues. The use of these would imply that the subject is in a way awakened by a signal which he is able to recognize and locate, but the variety of times chosen and the number of different tests carried out would make such coincidences difficult. What noises are there at 2:00 or 3:00 in the morning? Apart from exceptional cases it would be difficult to say. Furthermore, these authors have noticed that automatic waking takes place suddenly and abruptly with no transitional period, as if there were some disturbance in the room. It has even happened that the subject does not remember immediately upon waking that he had decided to wake up at that time. For all these reasons the use of external cues is not enough to explain these cases of spontaneous waking at a fixed time.

If the cues are not external, we must assume that they are internal. The organic cycles induced by circadian rhythms of activity, of which we have seen some examples, leave us in no doubt as to the existence of a physiological clock. But how can we *read* it? This problem brings us back to that of the nature of sleep: Can we be selectively conscious of certain signals during sleep? We know that the answer is yes. A mother hears the whimpers of her baby when the neighbours do not; the night watchman is wakened by a creaking door which he would not notice if he were not on duty. Sleep does not entirely suppress the function of vigilance, and psychoanalytical theory tells us the functional significance of dreams. It has been rightly noted that subjects who must wake up at a fixed time do not sleep as well as usual (more restless, quicker pulse). They behave as if they had something on their minds and this affects their dreams. The problem of the time and the worry of possible delays preoccupy them and they often wake up as the result of a dream (Bond, 1929). Thus dreams prove that the subject continues to "keep an eye on the time" while sleeping.

The time of waking can be determined by suppressed desires as well as by conscious will. Odier (1946) describes the case of a man who could wake up at precisely any time he fixed. One day, however, he woke earlier. An analysis of the situation showed that he had wakened at an hour which would have enabled him to go away with a woman with whom he was carrying on an illicit affair. This desire had been suppressed, but despite this it was responsible for his waking at a time when he could still have gone with her.

This concept of spontaneous waking does not imply some obscure intuition of time but simply the accurate interpretation of organic signals. This is an acquired ability for which children have far less disposition than adults (Clauser, 1954) and it also explains why automatic waking is more difficult the further the fixed time is from the normal time of waking. The signals are then less well known and more difficult to interpret than those which we have learned to locate in time through more frequent use. Moreover, in the middle of our sleep our vigilance is at its lowest and interpretation is more difficult.

It could be objected that these signals to which we are attaching such importance must be very weak since we do not notice them when we are awake. The study of dreams has shown, however, that, when asleep, we are sensitive to imperceptible organic stimuli; this explains in particular how we can have premonitory dreams of certain illnesses before the symptoms appear at a conscious level (Piéron and Vaschide, 1901).

Differences which occur from one individual to another do not affect the validity of these conclusions. They merely show that the relationship between vigilance and sensory cues is not the same in everyone. It should be possible to determine the exact typological characteristics of people who have this ability, and this would in turn explain the actual mechanism of temporal orientation by means of organic cues. Clauser (1954) is the only person who has attempted this analysis, but he based it on the classification made by Jaensch, which is of the most doubtful nature. According to him, most of the individuals who are able to wake themselves up at a fixed time belong to the "non-integrated" category (see p. 235) and in particular have a tendency toward dual personality and dissociation.

To conclude, we have seen that when organisms are exposed to periodic changes, they are usually able to adapt to them. In other words the variations in their physiological life synchronize with external changes and they learn to repeat beneficial behavior in anticipation of the periodic return of corresponding situations.

This adaptation to cyclic changes has an obvious biological significance, but it is also of psychological importance. The induction of organic rhythms from cosmic rhythms results in a double system of signals which are coordinated with each other. The preponderance of the cues provided by natural and man-made clocks obscures the importance of the acquisition of a physiological clock. Once the relationship has been established between these two types of signal, however, we can interpret organic cues whose temporal significance would not be at all apparent on its own. This probably explains a number of phenomena attributed to a so-called *time sense:* intuitive knowledge of the time, or certain cases of temporal orientation which are due more to physiological signals than to external cues. From this point of view, the ability of a man to wake himself up at a fixed time is no more mysterious than the spontaneous behavior of an animal at regular times, even in the absence of the external stimuli which originally determined it.

These adaptations are classical cases of conditioning, since the stimuli, which at first have no effect, subsequently become the signals for certain behavior. These cases are different, however, in that the unconditioned stimulus of external origin gives rise, by induction, to an associated internal stimulus. This induction determines the temporal location of the conditioned stimuli as a result of the synchronization between the external and internal series of changes. In this sense there is quite literally conditioning to time.

II

CONDITIONING TO DURATION

PERIODIC CHANGES ARE ONLY ONE PARTICULAR MANIFESTATION OF the universal state of change, and at that one of the most straightforward. They show succession and duration still in an elementary form. The phases of succession are always more or less repetitions: day and night, ebb and flow, etc. The length of time between the phases is identical, or at least the duration of a complete cycle is constant. What happens when men or animals have to adapt to changes which are not quite so simple?

I *Trace Conditioning*

We know that man has the capacity of apprehending succession and duration, but animals also have something more than a set of tropisms or simple reflexes which would only permit them to respond to present stimuli. Their behavior at any given moment takes into account what has already happened and what is about to follow; in other words, it allows for the succession of events. This is already apparent from conditioned reflexes. One stimulus becomes the signal for another which has the property of eliciting an innate or acquired reaction; when the conditioning process is complete, this signal will bring about the reaction on its own, even if, for some reason or other, the stimulus to which it responds is missing. Thus a form of training takes place which is basically a question of the succession of two stimuli. Pavlov always stressed the fact that no conditioning is possible unless the conditioned stimulus precedes the unconditioned stimulus, that is unless there is a succession (*Conditioned reflexes*, p. 27). Backward conditioning, when the conditioned stimulus follows the unconditioned stimulus, is impossible. This statement has been questioned, but the facts put forward as proof of the contrary could all possibly be explained by inaccurate control of the stimuli (Woodworth, *Experimental psychology*, pp. 121-122). The simplest form of adaptation to a succession of changes is seen when there is the least conditioning. A baby will stop crying when it sees its bottle, thus anticipating what is about to follow, just as a child a few months older will help his mother to dress him by holding out the appropriate arm or leg in turn.

It is not only the succession which counts; the interval between the stimuli also plays a part. To begin with, there is an optimum duration for this interval. Pavlov made no definite calculations but he stated that conditioning is established normally when the conditioned stimulus precedes the unconditioned stimulus by a large fraction of a second or by several seconds. Within this margin it is possible to *perceive* succession. However, conditioning, i.e., the ultimate establishment of an association between the stimuli, is most easily achieved when one follows the other at an interval of about half or three-quarters of a second. We shall see in Chapter 5 that this duration corresponds to the so-called indifference interval, which is the spacing of the stimuli at which perception of their succession is easiest: the two stimuli are distinct without appearing to be separated by any duration of time. This optimum interval will be found to be the same for different instances of conditioning involving the most varied responses: withdrawal of the hand (Wolfle, 1932; Spooner and Kellogg, 1947), blink reflex (Kimble, 1947; McAllister, 1953), and even the psychogalvanic reflex, although this is a reaction of the autonomic type (White and Schlosberg, 1952). Later we shall return to the fact that adaptation to succession is facilitated by perception of the latter; this will be seen more clearly in Part II.

Conditioning is still possible, although more difficult, when a noticeable interval separates the two stimuli. Moreover, in such cases the conditioned reaction takes place in the absence of the unconditioned stimulus after the same interval of time as that which separated the two stimuli during conditioning. In this case there is a double conditioned stimulus: the signal stimulus and the interval between the two stimuli. This is known as *delayed* conditioning. This brings about an adaptation not only to succession but also to duration, as the result of time judgment which we can now see is not limited to cases of periodic change. Thanks to delayed conditioning, time, in combination with conditioned stimuli, becomes a determining factor for behavior. These facts were proved in Pavlov's laboratory between 1907 and 1911 (according to Dmitriev and Kochigina, 1959).

This conditioning takes two different forms. The conditioned stimulus may be fairly long and the unconditioned stimulus take place toward the end of it; this is called *deferred* conditioning. On the other hand the conditioned stimulus may be short and precede the unconditioned stimulus by a long or short interval; this is known as *trace* conditioning, to emphasize the fact that the conditioned response is set off not by a perceived stimulus but by its memory trace (Pavlov, *Conditioned reflexes*, p. 39).

The laws governing the phenomena in these two cases are very similar.[1] The essential fact is that, after conditioning, the conditioned reaction is preceded by a latent period of the same duration as the interval from the beginning of the conditioned stimulus until the unconditioned stimulus.

It is even possible to establish a double delayed conditioned reflex to one stimulus in one animal. We merely condition the

---

[1] The laws are similar, but deferred conditioning is easier to establish than trace conditioning. Pavlov (*ibid.*, p. 92), notes: ". . . that the delay for a stimulus of one and the same character develops at a different rate according as the stimulus is continuous or intermittent. In the former case the development of delay is more rapid." Mowrer and Lamoreaux (1942) working under special conditions, observed the same fact in the case of a rat, while Rodnick (1937*a*) found the same for humans when conditioning a psychogalvanic reflex.

The fact has not been explained. We are tempted to compare it with the results we obtained when testing the appreciation of time in young children (Fraisse, 1948). "Filled" time has more reality than "empty" time and is judged far more accurately, as if the physical duration of the stimulus added another cue to those provided by "internal" processes.

short delayed reaction (15 seconds) in one room and the long delayed reaction (50 seconds) in another. The process is still difficult, however, in the case of a dog and may lead to neuroses. However, this experiment does show the complexity of the adaptations corresponding to these processes (Chu-Tsi-Tsiao, 1959).

We can follow up these points by considering the work carried out by a pupil of Pavlov's. In this case the conditioned stimulus was a whistle and the unconditioned stimulus of salivation an acid which followed after an interval of 3 minutes. If the number of drops of saliva produced by the dog is checked every half minute, the following results are found, after conditioning (Pavlov, *ibid.*, p. 90):

Time of Experiment	Number of Drops of Saliva
3.13	0 0 2 2 4 4
3.15	0 0 4 3 6 6
3.40	0 0 2 2 3 6

This example shows that response does not follow the conditioned stimulus immediately, but is *delayed.*

In this case the delay is not equal to the interval between the conditioned and unconditioned stimuli, or more precisely the reaction begins to take place before the end of the interval. This does not mean that the "estimation" of its duration is inaccurate. The delay in conditioning of this kind only gradually becomes apparent and Pavlov insists that the conditioning itself must be undertaken very slowly, although even then it is not always successful. Moreover, the salivary reaction is a *preparatory* response before eating and it is, therefore, normal for it to anticipate the appearance of the unconditioned stimulus. (Guillaume, *La formation des habitudes,* rev. ed., 1947, p. 33.) This is even more true when the conditioned reflex is of a defensive nature. Rodnick (1937a) found that the delay in the case of a psychogalvanic reflex set off by an electric shock preceded by the turning on of a light was 5.7 seconds after training, whereas the light was turned on 20 seconds before the shock (deferred conditioning).[2] The reflex linked with a defensive attitude on the part of the subject, or to an "anxiety," does not occur immediately the light is turned on but it precedes the shock by a fair length of time.

Would we obtain the same sort of results if we were to experiment on reflexes with a different biological significance? It is difficult to answer this question because most reflexes are of a defensive nature or are very closely linked with preparations for some activity. If the time of training is increased, however, experiments show that the delay in the reaction increases slightly; this increase is very precarious, however, for it drops again if 24 hours elapse between two sessions.

In one case, however, it did prove possible to show that the delay in the reaction is practically the same as the interval between the conditioned and unconditioned stimuli, although this was a physiological reaction and not one of behavior. Jasper and Shagass (1941) managed to establish trace conditioning of the disappearance of the cortical alpha rhythm, for which the unconditioned stimulus was a light. The conditioned stimulus was a sound preceding the light by 9.4 seconds. When the conditioning was stable, they measured the delay six times on 10 subjects and found a mean of 8.2 seconds with extremes of 7.2 and 9.2 seconds and a σ of 0.7 seconds. Even in this case, therefore, the delay is slightly smaller than the time interval between the two stimuli, but it is very stable and relatively accurate. We should note in fact that this biological estimation of delay in a physiological reaction was far more accurate and less variable than the conscious and voluntary estimation made by the same subjects. They were asked to estimate the interval between the conditioned and unconditioned stimuli by pressing on a key when they thought the latter should be due. Their estimations varied from 6 to 15.2

[2] The delay is only 4.1 seconds after a very long training in the case of a conditioned *trace* reflex, i.e., when a brief flash of light preceded the shock by 20 seconds. Obviously these figures have only a relative value. In the case of deferred conditioning, for instance, Switzer (1934) carried out an experiment similar to Rodnick's and found that the delay increased from 5 to about 10 seconds for a shock applied 16 seconds after the light was turned on.

seconds with a σ of 2.5 seconds. There is no correlation between the physiological delay and the conscious estimation, which proves that the two processes are independent.

Trace conditioning, then, shows that individuals take duration into account at an involuntary level in their adaptations to their surroundings. It would even seem from Jasper and Shagass' experiment that the physiological registering of duration is quite accurate and that it is other psychological factors which often cause anticipation of the reaction.

II *The Estimation of Duration in Operant Conditioning*

There is a real difference betwen the classical type of conditioning and operant conditioning (Hilgard and Marquis, *Conditioning and learning,* 1940). In the latter case, instead of being given an unconditioned stimulus, the animal has to learn to react in a certain way in order to avoid pain or obtain satisfaction. Repetition is still necessary for the conditioning process, so that the right response can be found by experience and retained by reinforcement. Those techniques used for operant conditioning also make it possible to see that animals can perceive duration; they provide in particular a means of determining the precision of what we shall call, anthropomorphically, the estimation of time by the animals.

When put in a maze, a rat will choose the shorter of two possible routes; similarly, if there is a choice between two confinements of different duration, it has been shown by experiment that it will choose the shorter. This was proved by Sams and Tolman (1935). When the rats entered the apparatus they had to choose between two passages which were identical from every point of view; every time they went through these they were detained in a small chamber, 1 minute in one passage and 6 minutes in the other, before they could reach their food. Gradually they learned to choose the chamber where they were kept only 1 minute. This was not a spatial preference, because when the duration of their confinement in the two chambers was reversed, the animals also reversed their choice. By this method it is possible to judge the capacity of animals for discriminating between two intervals of time.[3] A. C. Anderson (1932) used it systematically, giving the animal a choice between four durations of detention. His rats passed 500 times through the apparatus in the course of more than 3 months; at the end of this time, in 72 percent of the cases they went through the passage where they were detained for 1 minute, in 19 percent through the passage where they were detained for 2 minutes, in 6 percent through that of 3 minutes and in 3 percent through that of 4 minutes. Learning is thus unquestionable. Using the same method, but with only two passages, the same author established that the rats were more sensitive to the relative difference between intervals of time than to their absolute difference; this shows that Weber's law applies to time discrimination in animals.

We have drawn up the following table from Anderson's results:

Relationship Between the Durations				4/1	3/1	2/1	1.5/1
Percentage Choosing the Shortest Detention				96	84	76	65

Difference Between Durations							
4 min.	3 min.	2 min.	1 min.	30 sec.	20 sec.	10 sec.	
Percentage Choosing the Shortest Detention							
82	96	77	79	71	80	74	

These values are only indicative because the two groups of rats cannot be compared. They show, however, that the percentage of successes decreases regularly as the relationship between the durations decreases, whereas there is no law for absolute differences even though large differences correspond to large relationships.

But is this really temporal discrimination? Hull (*Principles of behavior*, 1943) points out that the interpretation of these results

[3] Woodrow (1928) used a somewhat similar method in his study of time discrimination in the *monkey*. The monkeys were allowed food after distinguishing the longer of two empty intervals given one after the other at random. He found that these monkeys could discriminate between 1.5 and 2.25 seconds. Being within the limits of perceived duration, these intervals come under the heading of another type of behavior (see Part II); we only mention them here for interest's sake.

is ambiguous. Does the duration of detention act as a differentiation cue, or does the animal choose the route with the shortest confinement simply because its choice is more quickly reinforced? Obviously this second factor may play some part, because the more quickly the reward follows the choice, the more readily the latter is fixed. Cowles and Finan (1941) proposed another method whereby the interval of confinement preceded the choice instead of following it and was thus the only basis for discrimination.[4]

They used a Y-shaped maze. The animal first entered the common passage, which formed the detention chamber. It then had to choose the left or the right hand branch according to whether the confinement was long or short, only one choice being rewarded. By this method, time is the only discriminating factor. This training proved to be possible, thus demonstrating once again that animals are capable of time discrimination. It seemed more difficult, however, than in experiments of the type carried out by Sams and Tolman. Six of the nine rats of Cowles and Finan learned, after 600 trials, to discriminate between 10 seconds and 30 seconds with 70 percent success. Heron (1949), who used an apparatus constructed on the same principle, found very marked individual differences among the rats. Of 11 animals, three could only distinguish between 5 and 45 seconds, four between 5 and 25 seconds, three between 5 and 20, and only one between 5 and 10 seconds.

These methods prove that animals are capable of adaptation to duration and of discrimination, but they are not suitable for quick or precise determination of the accuracy with which subjects can estimate duration. Ruch (1931) suggested another method. An animal is placed on a grid A; to reach its food it must pass over another grid B and through a door C. At first B is electrified and C locked; if the animal gets on to B it can only

[4] Mori (1954) also tried to check the validity of Hull's objection, but his technique was not so effective. He varied the location of the confinement chambers, bringing them nearer to or further away from the goal, but could observe no influence of this variable on the behavior of the rats. He also tried putting two chambers instead of one in each branch of the maze. On one side the rat was detained for 1 minute then 7, and on the other side 7 then 1. The rats did not seem to be capable of perceiving any difference between these two situations.

return to A. After a certain length of time B ceases to be electrified and C is unlocked but remains closed. The animal must take advantage of this moment to cross B and go through C, for if it stays on A it will receive an electric shock. To solve this problem, therefore, it must estimate a length of time shorter than that between the beginning of the experiment and the point when the grid A is electrified but longer than that during which B is electrified. These upper and lower limits define a period of safety which the experimenter can vary around a mean value. The training is fairly quick and the safety period can gradually be shortened. If we then calculate the difference between the time of the first attempt to leave A and the middle point of the safety period, we find, in the case of a rat, an error of 13 percent which can be considered as a threshold value determined by the method of the mean error. This threshold value is very small: for approximately the same duration (in this case 438 seconds) the threshold value for humans is around 20–30 percent. Buytendijk (1935), who used a method similar to Ruch's, noted that anticipated reactions were more numerous than delayed reactions, as is the case for classical conditioning.[5]

It has thus been proved by very varied methods, that adaptation to time does take place in animals, where the complex processes that we shall study in subsequent chapters do not exist; either they submit to the law of time, as is very evident in the case of classical delayed conditioning, or their behavior is determined by time, as in instrumental conditioning.

It is true that the training is long and difficult and that individual differences are very great (Pavlov, 1929). It is also true that the result is very unstable; an interruption of 24 hours in training sessions leads to a reduction in the delay in the case of deferred or trace conditioning (Switzer, 1934; Rodnick, 1937a). A delayed response is also less intense than an immediate response (Kotake and Tagwa, 1951).

[5] Although it is comparatively easy to make rats estimate duration, it seems to be more difficult to make them respect it, for instance by rewarding them only if their responses (pressing on a lever) follow a signal after a certain delay. The experiment is possible with a delay of 10 seconds, but if this is increased the rats hardly adapt at all to the new situation (Wilson and Keller, 1935).

It also seems that trace conditioning is only possible within fairly narrow temporal limits, although this question has not been thoroughly studied. If the delay is too long, a connection is no longer established between the signal stimulus and the response, and discrimination between two durations is no longer possible.

III Psychophysiological Interpretations

How can we explain the mechanism of this temporal adaptation in the case of animals, where we cannot attribute anything to the higher processes which we shall be analyzing in Part III?

Pavlov (*Conditioned reflexes*, p. 92) thinks that the conditioned stimulus gives rise at first to a conditioned inhibition (inactive phase), followed by a period of excitation (active phase). To prove this, he points to the fact that the phenomena observed during the first part of the period of delay correspond in every way to the processes of inhibition. In particular, any stimulus of average strength is enough to elicit salivation in a dog in the first phase by ending the inhibition; on the other hand if applied at the end, during the phase of excitation, a stimulus of the same kind will have an inhibiting effect and thereby reduce the salivation.

Let us take an example which shows the ending of inhibition by a neutral stimulus. The conditioned stimulus is mechanical excitation of the skin and the unconditioned stimulus, which follows after 3 minutes, is an acid. The salivary reaction is delayed (experiments carried out from 9:50 to 10:30). If the noise of a metronome, which has had no relationship with the salivation, is now associated with the conditioned stimulus, salivation will be seen to begin immediately (10:15). Pavlov gives the following table (*ibid.*, p. 93):

Time	Stimulus	Number of Drops of Saliva Every Half Minute					
9.50	tactile	0	0	3	7	11	19
10.15	tactile + metronome	4	7	7	3	5	9
10.30	tactile	0	0	0	3	12	14

It seems that the metronome, which had no stimulus value for the dog, ended inhibition in the first phase and created it in the second; this is a characteristic effect of a neutral stimulus acting in accordance with the state of inhibition or excitation of the brain. Pavlov also finds proof that delay is the result of inhibition in the fact that anything which hinders the inhibiting process makes it more difficult to establish trace conditioning. This can be seen in types of dogs which have weak inhibiting processes or simply in dogs whose excitation process is reinforced by lack of food before the experiment.[6]

This hypothesis has also been confirmed indirectly in several other ways. The fact that respiration is inhibited in man during trace conditioning of the psychogalvanic reflex could be given this interpretation (Switzer, 1934), although apprehension may be a sufficient explanation. By considering the reaction time as conditioning, Leridon and Le Ny (1955) showed that a subject who had got used to reacting to an unconditioned stimulus 5 seconds after the conditioned stimulus (or preparatory signal) showed an appreciably longer reaction time when the unconditioned stimulus followed after only 2 seconds, i.e., during the phase of inhibition. This result is particularly surprising in view of the fact that the reaction time is usually shorter after an interval of 2 seconds between the preparatory signal and the signal of execution than it is after 5 seconds.

According to Miller (1944), the existence of approach and avoidance gradients can also be interpreted in the same way. The drive grows gradually stronger as the moment for the expected stimulus grows nearer. Let us take Rigby's experiment (1954) as an example. The head of a rat is put in a harness, the movements of which will show its approach or avoidance behavior. For the approach situation, a light is put on 10 seconds before the appearance of food; for the avoidance situation, a noise is made 10 seconds before the rat is given an electric shock on its hind

[6] If we compare trace conditioning with expectant behavior, we can see, as Birman did (1953), that individuals who are able to bear waiting well are most likely those in whom the inhibiting phase predominates. On the other hand, the inhibiting process is weak when there are manifestations of impatience and general emotional reactions.

feet. After the training, which results in trace conditioning, the approach drive is measured at various points of time during the 10 seconds interval; it is found to increase slightly over the first 8 seconds, then suddenly toward the end. The avoidance drive follows the same pattern.

Evidence for the process of inhibition is perhaps also given by the fact that it becomes generalized during the delay. The conditioned stimulus of some other reaction has in fact little or no effect during the inactive phase (Koupalov and Pavlov, 1935). For instance, Rodnick (1937*b*) has shown that a blink reflex, conditioned by a noise, is less intense if the noise is heard during the trace conditioning of a psychogalvanic reflex.

It is therefore very likely that a process of inhibition develops during trace conditioning, but we still have to find the actual mechanism which controls its duration; for this we must pass from description to explanation.

Pavlov was content to class trace conditioning with the conditioning to periodic change which we studied in Chapter 1. This would mean that for each phase of the delay there is a corresponding state of the organism that brings about a process of cortical inhibition during the delay and a process of cortical excitation at the moment of the reaction. But what are these successive states of the organism? If we feed a dog every 30 minutes, it is not unthinkable that there is a different state of the digestive process corresponding to every moment of these 30 minutes, but what should we think when a noise alone is the signal for food? We should have to postulate that a series of reactions develops in the organism; but, as in the case of time conditioning, we should have to explain the mechanism of the temporal regulation of these reactions when their succession is not governed by the laws of some biological process such as digestion. This brings us back to Popov's hypothesis of *cyclochronism:* the nervous system is capable of reproducing a series of excitations in the same order and at the same temporal intervals as when they first acted on the organism. This hypothesis can be applied equally well to periodic changes and to repetitions of identical sequences. The conditioned stimulus/ unconditioned stimulus group, including the interval between the two, constitutes one entity which tends to be reproduced identically. The pacemaker, of cyclochronic origin, does not exclude the creation of a phase of inhibition, but this is only an effect. It also does not exclude the possibility of the animal using other cues either from its surroundings or from the organism.

In the case of operant conditioning, several authors have been of the opinion that the animal may be guided by the amount of its own activity during confinement. Cowles and Finan (1941), however, observed that animals have no typical or regular activity during this period. Heron (1949), for his part, suggested that a certain tension develops while the animal is confined and, according to the level this has reached when it is released, it chooses one or the other solution. This hypothesis remains rather vague, but it may be corroborated by an observation made by Buytendijk (1935). He put a dog in a harness of the kind used by Pavlov and trained it to push open a flap, behind which a piece of meat was placed, every 90 seconds. Its respiration was recorded continuously by a pneumograph and showed a constant increase in volume as the moment of action approached. But which is the effect and which the cause?

The most probable explanation is that a specific nervous process acts as a pacemaker and that this process is brought about by the repetition in sequence of excitations which are of vital importance for the organism. The center which controls periodic adaptation and trace conditioning is probably subcortical. The research that is based on operant temporal conditioning seems to confirm the results obtained from periodic inductions (Chapter 1, pp. 28-39). Finan (1939) removed the frontal lobes of four monkeys and found that they were equally capable, before and after the operation, of solving a temporal problem of the kind set by Ruch (see p. 56); to stay on one grid for 10 seconds then cross over another grid during the next 10 seconds in order to avoid a shock from either of them. They are even able to discriminate between a confinement of 30 seconds and another of 120 seconds.[7]

[7] Using the same experiment, Finan showed that these same animals were incapable of delayed responses. A delayed response is one in which a delay is imposed between the signal for the reaction and the moment of response.

When taken by itself, this experiment only proves that the frontal lobes do not have to be intact for trace conditioning, but, as it is considered that the regulation of temporally organized behavior may be located in these lobes, it may be inferred that the basic pacemaker is most likely subcortical. Obviously this center, which would register duration and provide cues for temporal sequences, could only act in relation to the whole behavior of the individual. Thus the laws of adaptive reactions, which imply anticipation of the situation, may explain the fact that in the case of trace conditioning of a psychogalvanic reflex or a salivary reflex the delay is less than the interval between the conditioned and unconditioned stimuli. The same may be noticed in the case of anxiety reactions.

This combination of physiological processes and cues provided by emotion or by action also explains why time estimation is less accurate the more complex the conditions of adaptation. This paradox simply means that only the basic physiological process gives us an accurate cue; it is confirmed by the fact that trace conditioning of the alpha rhythm, which is not connected with behavior, is the most accurate and the most regular of all the cases of conditioning studied.

In conclusion we must consider what part conditioning to duration plays in the life of man. We have seen that animals are able to make practical estimates of duration, although they have no symbolic representations of this and are not capable of any intellectual processes. Does this primitive ability still exist in man?

Thus if a piece of fruit is hidden in front of a monkey, there is a positive delayed response if it manages to find it after an interval. Jacobsen (1936) found specific disturbances in these responses in animals whose frontal lobes had been removed. Finan (1939, 1942) compared the results obtained with such animals in strictly temporal experiments with their failures in the field of delayed reactions and thus showed clearly that there is no truly temporal factor in delayed responses, although they are often classed with trace conditioning. The failure concerns memory and is due to a lack of fixation of the stimulus (hiding place of the object). Time plays a part here, not because of its duration—which does not need to be registered—but because it makes it possible for the animal to forget between being shown the object and having to remember it.

It is not immediately apparent, but this is not surprising, for if it does exist it is only manifested in connection with the more complex forms of behavior which are not observed in animals. However, the similarities we have seen between adaptation to periodic change and conditioning to duration give us reason to suppose that man too is capable of registering duration on a biological level. In the case of temporal orientation, we have clearly seen that man uses both the organic cues provided by the adaptation of his body to change and the most elaborate symbolic representations of time; in the same way, promptings of biological origin must combine with constructions of the mind for the estimation of duration. This is perhaps why the direct estimation of time still keeps its air of mystery and often surprises us with its accuracy.

III

THE PERCEPTION OF TIME

IIIIIIIIIIIIIIIIIIIIIIIIIIIIIIIIIIIIIIIIIIIIIIIIIIIIIIIIIIIIIIIIIIIIIIIIIIIIIIII

While we are borne along and shaped by universal change, we can also witness its happening, for we perceive it as change. This perception is possible in so far as we can apprehend with relative simultaneity several successive phases of change which thus appear to be linked.

Consequently, the perceived present has a temporal extension, the duration of which is bound by the limits of the organization of successive elements into one unit (Chapter 3).

The threshold of the perception of succession—the moment when two stimuli cease to appear simultaneous, and the threshold of duration—beyond the instantaneous, vary according to conditions and the sensory receptor (Chapter 4).

Perceived duration is only one characteristic of the organization of succession; its quality depends on the speed of succession of the stimuli and its quantity on their nature (Chapter 5).

As far as we can tell, the higher vertebrates perceive time under much the same conditions as man. This form of adaptation, therefore, appears to be closely linked with the general properties of the receptor centers. In the case of man, however, perception is not only a guide for his immediate reactions but also a source of knowledge.

||||||||||||||||||||||||||||||||||||||||||||||||||||||||||||||||||||||||

THE PSYCHOLOGICAL PRESENT

IN SOME CASES WE NOTICE THAT A CHANGE HAS TAKEN PLACE, while in others we perceive it happening. Not long ago the sun was shining; now it is overcast. I was absorbed in my work and did not perceive the change, but I notice it now thanks to my memory. On the other hand, the bell of the nearby school is ringing at the moment and I can perceive the succession of these brief alternations of sound and silence which correspond to the striking of the clapper. In this case I actually perceive the changes, as I could have perceived the clouds passing over the sun just now.

This chapter is devoted to a general analysis of the characteristics and conditions of our perception of change. But to define the problem we must first decide exactly what we mean by the process of perception so that we can distinguish this from other methods of adaptation.

We are always being let down by psychological terminology, which is better adapted to practical life than to a science. In everyday language the word *perception* means any awareness, whether the process is sensory, affective, mnemonic, or intellectual. To take one example which is appropriate in this context, Guitton (*Justification du temps*, 1941, p. 19) speaks of "perception of the future." In this case the word is obviously used metaphorically. To perceive—we shall use the verb, since perception is the action of a subject—implies essentially a reaction to the *present* situation. This situation may be described in physiological terms as the existence of an excitation of peripheral origin in the nervous centers, in terms of consciousness as an immediate experience, and in terms of behavior as a reaction adequate to the situation. From all these points of view, we can speak of perception only if there is an immediate response to a present stimulus. We could perhaps define the word even better by distinguishing it from what it is not. Remembering is the recalling in deed or in thought of a stimulus which is no longer part of the present experience of the subject. Thinking is the relating of perceived, remembered, or imagined events. Affective reaction is the effect, shown in the entire psychic activity, of an occurrence which is again perceived or called to mind. Before any of these processes can take place there must of necessity be a direct or immediate experience.

No confusion at all is possible between perceived space, imagined or recalled space, and space conceived as an abstract idea. In a different context, Michotte speaks of "perception" of causality, because he has shown that by modifying the information within the present field of perception of the subject, he could at the same time modify the latter's reactions; in his experiments these were verbal. The fact that perception is a reaction to a present occurrence does not necessarily mean that it is independent of any previous experience. In the course of the lifetime of any individual, his sensory information is gradually enriched with significance acquired by conditioning. A dog salivates at the sight of meat, but only if it has eaten meat before. I say that I perceive a sheet of paper because since my childhood the word *paper* has always been associated with an original experience. Thus the present is full of significance acquired in the past, but it is a different thing when the past is remembered or "relived" through a present stimulus. The nature of any perception depends to a great extent on the past experience of the individual, but it can always be defined as the apprehension of present stimuli without the explicit intervention of memories and without intellectual elaboration.

From this it is immediately clear that the perception of change poses a problem. To speak of change is to say that what is, ceases to be or is transformed. Whether I note the change by comparing what I perceive with what I remember, or can grasp it by relating these successive glimpses of the world, the case is clear. But do we only apprehend the world as a series of tableaux rather like a child's picture book of stories? Actually we know that our perceptions are not static and that we *perceive* a world in a state of perpetual change.

We must distinguish between continuous change and discontinuous change. Continuous changes result in the perception of a *transformation* in quality or intensity: the darkening of the sky or the spreading glow of the sunset, the growing noise as a car approaches. A change in place is perceived as *movement*. This kind of change is always characterized by its speed, and speed is a perceived fact sooner than a concept which can be defined by measurement (quantity of change per unit of time). A five-year-old child who is not capable of measuring, and does not even possess the concept of speed, can still compare the speed of two moving objects, provided that the situation is not so ambiguous that it requires mathematical calculation (Fraisse and Vautrey, 1952). These changes are perceptible only if their speed reaches a certain threshold value. Below that we notice different, successive states of the same phenomenon, but we do not see the transformation. This is the case for the growth of plants. It has taken the cinecamera and the possibility of projecting films faster than they were shot to let us *perceive* the growth of a stem or the bursting of a bud.

The perception of continuous transformation gives us rather indistinct temporal information because we can only perceive clearly that which is static. It permits us, however, to anticipate the *successive* states or positions of whatever is changing. The amateur photographer who develops his own film can see the negative change more or less quickly, and from this speed he can tell the right moment to take it out of the developer. The perception of speed also permits us to anticipate the moment when a moving object will pass a fixed point; thus the pedestrian knows when it is safe to cross the road and the hunter where to aim his gun to hit a hare or a partridge.

Through these transformations we can experience what is

durable; they suggest this to us by the constant renewal of their evanescence. But we cannot perceive in this any quantitative duration. The latter is always an interval between two phenomena or two distinct states of one and the same phenomenon.[1] The river which flows beside me is fast or slow, but its movement does not give me any perception of duration. From speed we can of course deduce duration, but only by a process of reasoning. Nothing that we perceive tells us that what goes faster does not last so long (see Chapter 8).

The perception of continuous change has never been doubted. The perception of discontinuous change, on the other hand, raises some very complex problems. Continuous change can be perceived at any given moment, whereas if I perceive discontinuous change it implies that I perceive not only state A and state B but also the transition from one to the other, or their succession. To speak of succession is to say that when a new phenomenon occurs, the old one is no longer present. At least this is the analysis prompted by our *idea* of succession. If we define perception as the apprehension of the present, perception of a succession seems impossible; this is the conclusion reached by *logical* analysis, which might be reinforced by hasty introspection. The Würzburg school has fortunately put us on our guard against objective error, which consists in confusing the phenomenon observed with our knowledge of the object.

Bergson made the subtle observation that perception of succession implies the simultaneous, not successive, perception of *before* and *after*, and that it is "paradoxical to assume a succession which is nothing but succession but which nevertheless

[1] This observation is probably at the root of Bergson's idea of making a distinction between quality duration, which he considers as true duration, and quantity duration: "Pure duration is what the succession of our states of consciousness becomes when our ego drifts through life and refrains from drawing a distinction between the present state and previous states." (*Essai sur les données immédiates de la conscience*, 19th ed., 1920, p. 75.) Bergson refers here to a special change, that of our states of consciousness, and shows in this an experience where our thoughts and even more our emotions fuse together in perfect harmony. This experience is not allied to perception, because if we tried to find the exact stimuli corresponding to it we would, by this very act, destroy the state of fusion. We shall be discussing Bergson's attitude again later (see p. 79).

takes place in a single moment." (Bergson, *ibid.*, p. 77.) We know Bergson thought that we apprehend succession only because we project it in space and make the end of the first state coincide with the beginning of the second. This *representation* of succession actually corresponds to a type of behavior which we shall study in Chapter 8. It is possible, however, to prove that we can perceive succession.

The first stage of our proof must remain on the level of phenomenology for we must show that the paradox of the simultaneous perception of before and after exists only on a logical level and not on the factual level. First we should stress that not all discontinuous changes are perceptible. It is 3:15: the clock has just struck once. As far as I can perceive, this single chime does not succeed another; but my memory tells me that it follows on the three chimes that struck 15 minutes ago. On the other hand, when the clock strikes four at the next hour I shall perceive the succession of these four chimes; this means that the first chime will still be present in some way when the fourth strikes. There is nothing paradoxical about this if it is true, as we shall show, that the perception of a stimulus, however brief, is an action which itself has a certain duration and, as it were, creates a bridge between physically successive stimuli.

In cases where there is perception of succession, two situations may arise. If the stimuli are numerous, rapid, and regular we perceive succession, but the predominating feature of this succession is its frequency. If the frequency is high, the individual elements are indistinct and we only have the impression of a change in the intensity of one and the same stimulus (tactile vibrations, flicker, crepitus). If it is lower, we perceive a succession of separate elements, such as in the quick beat of a metronome. The perception of frequency is allied to that of the velocity of a continuous change and shows us indirectly that the latter type is actually perception of multiplicity. All a high-speed camera does is catch one static state of a phenomenon. The perception of change is always multiple apprehension. Frequency, like velocity, gives us information concerning the quality of a change but not concerning its duration.

If the frequency is fairly low (2 or 3 to the second), and we pay attention for a certain length of time to the sequence of the stimuli, we see a new aspect of the perception of succession. Let us take sounds as an example: if we listen to a succession of identical sounds following each other at regular intervals, they will seem to us to be grouped in twos or threes; to our perception they are no longer individual sounds but successive groups. Through this phenomenon, which has been called "subjective patterning" to emphasize the fact that it does not correspond to a physical reality, more than through any other, we are able to analyze the perception of succession. This grouping is the result of a comprehensive and, as it were, simultaneous apprehension of several elements which form one unit of perception. This unit is not only observed introspectively. It also has an effect on our behavior. The year-old baby who rocks in time to music and the schoolboy who does the same while reciting poetry show that the line or the bar has unity for them. If movements are synchronized to sounds that recur periodically, the subjective patterning is shown by cyclic variations in the quality of the movements (Fraisse, *Les structures rythmiques,* 1956, p. 20).

Rhythm is not an unusual phenomenon but a special case where the unit of successive elements is better seen because it is always repeated identically. The organization of successive elements into units of perception is, however, such a fundamental part of our experience that we no longer notice it. It is the basis of our perception of rhythm, of melody, and even of the sounds of speech. A baby learns to attach a meaning to the phonemes *ma-ma* only because it perceives them as a unit.

Since the perception of succession in units is a basic feature of the perception of time, we shall analyze its characteristics successively and show that it is: (1) perception of order and (2) perception of the temporal interval. We shall then (3) determine the limits within which it is possible.

I *The Perception of Order*

Let us take a very simple case: the sound of a clock. I perceive a "tick-tock," then it is gone and another "tick-tock" takes its place. When the second "tick-tock" is heard, the first is no longer present and only memory, an immediate memory, permits me to know that this "tick-tock" was preceded by another. But when I perceive a "tick-tock," the "tick" is not yet part of my past when I hear the "tock." Thus I perceive directly the order of the "tick" and the "tock" without even having the idea of succession and without the help of my memory.

Nevertheless, the perception of order is only possible if the successive stimuli can be organized among themselves; that is, if they are of the same nature. A succession of sounds and lights will never allow perception of an organization which integrates the two. There will be perception of a double series, one of sounds and one of lights. This is what usually happens in choral singing, where each part has its own individual organization. In the next chapter we shall also see that it is difficult to perceive the order of succession (or the simultaneity) of two dissimilar stimuli, such as a light and a sound, because they do not fall into a spontaneous order.

The work of Broadbent (1958) has added an important contribution to our knowledge in this sphere. If three digits such as 7–2–3, are presented visually and at the same time three others, e.g., 9–4–5, are spoken, the subject can only reproduce these stimuli by repeating the auditory series after the visual series, or vice versa. The visual stimuli have become organized among themselves and so have the auditory stimuli. This phenomenon of grouping can also be based on the place of reception; this is the case when one set of stimuli is presented to one ear at the same time as another set is heard by the other. The same law also applies when two voices are heard simultaneously on two different wave lengths. In every case, therefore, the grouping takes place on the basis of qualitative similarity of the stimuli.

It is important to note that the order of succession is *perceived*. It is not the result of an organization imposed on stimuli which are independent of each other, as if we were threading beads. In the organization of perception (spatial or temporal), the activity of our minds does not impose a form on dissimilar elements, for in every field of our knowledge, order has its own laws and cannot be forced (Guillaume, *Introduction à la psy-*

chologie, 1949, pp. 339-340). Furthermore, we do not grasp this order by studying a "representation" of successive events. If this were so, having perceived three successive elements A—B—C, we could take them just as easily in the order A—B—C, C—B—A, or B—A—C, etc. But this is not so. It is easy to reproduce digits in the order in which we have heard them; in fact, this is the spontaneous attitude found in even the youngest subjects. It is far more difficult to reproduce them in any other order; to do so, we have to fall back on the intermediary of a representation. The order is inherent in the stimuli themselves and in the case of rhythm it is practically impossible to reproduce the individual elements in any other order.

Many theorists assert that memory is the explanation for our experience of order, which is of fundamental importance in the whole psychology of time. In the nineteenth century, the associationists tried to determine how we could take a multiplicity of sensations and establish their order, thus experiencing change and knowing their position in time. The same problem arose for space. To account for extension, Lotze proposed the theory of local signs (*Medizinische Psychologie,* 1852; quoted by Ribot, *La psychologie allemande contemporaine,* 1879). As every inch of our body, for touch, and every point on our retina, for vision, has different sensory receptors, we have sensations of differing intensity for the same stimulus and this provides us with a *local sign.* The movement of the body (or of the eyes) permits us to discern the local signs for different points in space. Thus the possibility of space results from the organization of these signs, through an innate idea of space; in this Lotze was indebted to Kant. As regards time, many authors have speculated on the nature of *temporal signs:* alternating sensations of tension and relaxation may impose a sign on successive sensations and thus make it possible to give them an order. At that time most authors sought these indications in the degree of obliteration of our sensations. Lipps, for instance, says that sensations are born and then obliterated: the degree of obliteration of two sensations at a given moment corresponds to their position in time. The difference in intensity of images gives their difference in position

(*Grundtatsachen des Seelenlebens,* 1883, quoted by Bourdon, 1907).

The most elaborate theory is decidedly that of Guyau: originally there was neither coexisténce nor succession, only a multiplicity of sensations and images, each different from the others. "Moreover, even memory has its degrees according to whether it is near or distant; any change which is registered in our consciousness leaves there, as a residue, a series of images arranged in a sort of line in which all distant images tend to become obliterated to leave room for other, more distinct images." (Guyau, *La genèse de l'idée de temps,* 2nd ed., 1902, pp. 25-26.) He adds that "the proof that the representation of before and after is a play of images and residues, is that we are very well able to confuse them" (*ibid.,* p. 26). As an example, he quotes the experiment in which a subject confuses the order of appearance of two sparks which glow at two different points in space. The one we look at appears to be the first. But Guyau has chosen one example in which there is no perceptual organization of the stimuli owing to their spatial discrepancy. According to him, order only results from a sort of sedimentation of our memories. There are objections to this theory: if order is related to the intensity of the images, how, asks Bourdon (1907), can we explain the fact that after hearing a series of letters, it is far more difficult to repeat them in reverse order than in the order in which we heard them? This reasoning is valid for obliteration as much as for representation. As the most recent elements are the least obliterated, they should be reproduced first and with greater ease; but the opposite is true. Furthermore, when the interval between two stimuli is very short, the difference in the degree of obliteration cannot be perceptible. "Generally speaking, it is hard to admit that there exist, between the images of successive impressions, differences in intensity comparable, in the fineness of the distinction, to the differences of position in time that we are *capable of perceiving.*" (Bourdon, *op. cit.,* pp. 474-475.)[2]

_____

[2] We shall deal with the problem of the memory of succession in more detail in Chapter 6, p. 159.

In actual fact, our perception of the order of sensations cannot be reduced to another mechanism. The order is given in the actual organization of succession. One condition is necessary, however: the organization must be *spontaneous* and this happens only if the stimuli are homogeneous and all take place within certain temporal limits. If the interval between the stimuli is long enough—we shall define this later—when the second stimulus occurs the first is no longer part of the perceptual present and no succession is perceived: the two stimuli are distinct and their order, which is not perceived, must be reconstituted by memory. In this case signs may be used, such as the degree of obliteration, and also logical reconstructions (Chapters 6, 8).

II *The Perception of Duration*

When we perceive succession, we apprehend at the same time an ordered multiplicity and the intervals separating the individual elements, that is to say, the durations. "Duration is, as it were, the measure of succession, the value of the interval." (Delacroix, *La conscience du temps*, 1936, p. 306.) We cannot isolate these durations or intervals except by analyzing our perception: we do not perceive duration independently of that which endures, just as perceived extension is always that of some object. "The phenomena of duration are constructed with rhythms; but the rhythms are by no means necessarily founded on a uniform and regular temporal basis." (Bachelard, *La dialectique de la durée*, 1936, p. 5.) In other words, the perception of duration is that of the duration of an organization.

It is possible to show objectively what we find by subjective analysis. When the organization is not obvious, it is difficult to perceive duration. For instance, if two adjacent temporal intervals are marked by a sound at each end and a light in between, i.e., by a sequence of sound-light-sound, they cannot be compared with anything like the same precision as two intervals marked by three identical sounds, for the latter form one unit of perception. In the case of the sequence sound-light-sound, a comparison is only possible at all because we impose on this heterogeneous sequence a homogeneous succession which permits of organization, such as three movements of the hand or three phonemes, each corresponding to one of the three stimuli (Fraisse, 1952a). Moreover, a succession is characterized not by its elements, either stimuli or intervals, but by the scheme of its durations: at the time when Von Ehrenfels was revealing the importance of formal qualities. Mach showed that the identity of the temporal patterns was what made it possible for rhythms composed of different *elements* to appear similar (Bouvier, *La pensée d'E. Mach,* 1923).

It can also be proved that perceived duration—like perceived extension—depends on organization, by comparing temporal illusions with spatial illusions.

Just as in Oppel's illusion a divided straight line appears longer than an undivided line, so an interrupted temporal interval seems longer than an empty interval (see Chapter 5, p. 132). Let us take another example: A basic law of perception is to minimize small differences (tendency toward assimilation) and exaggerate appreciable differences (tendency toward contrast). This law applies equally to the perception of spatial structures and temporal structures (Fraisse, 1938).

We can push the analogy between spatial and temporal structures even further by showing that the distinction between figure and ground can be applied to the latter. Let us return to the "tick-tock" of a clock. The "tick" and the "tock" are organized together and delimit an interval which has a duration. But between the "tock" and the "tick" of the next double sound there is another interval which is perceived only as a gap of no defined duration. The role of this interval is analogous to that of the *ground* of our spatial perceptions. The latter is in fact characterized by its lack of form, and the interval between two "tick-tocks" also lacks that form given by the organization of succession which is duration. This analysis may be confirmed by observations of behavior. If subjects listen to rhythmic structures repeated identically, they will be found to reproduce the intervals within the structure accurately but they cannot grasp spontaneously the duration of the interval between the rhythmic groups (Fraisse, *Les structures rythmiques,* 1956, p. 74).

Other experiments of the same kind, involving motor reproduction of auditory structures, help to show the functional difference which exists between the intervals within an organization and those which separate one organized group from another. If the duration of an interval within a rhythmic group is changed, the apparent duration of the other intervals is modified (as is the character of the whole unit). In other words, the modification of one part leads to a reorganization of the whole; this is also characteristic of figures in space. On the other hand, if the duration of the interval between the rhythmic groups is changed, the latter are not modified. From the point of view of perception, the interval between them is nonexistent—it is just a gap (Fraisse, *ibid.*, p. 73).

The importance of these gaps was formerly not appreciated, for it seemed that we *filled* them with our own duration; but Bachelard, contrary to Bergson, insisted that there is never at any level continuity of duration, but always an alternation of fullness and emptiness, of action and rest (Bachelard, *op. cit.*, p. 3). On the level of perception which we are at present discussing, these gaps play an important part for they keep the units of succession separate and this, in the case of speech for instance, permits them to become units of meaning.

So far we have dealt with intervals between successive stimuli; these are usually known as empty durations or even empty time, as opposed to filled duration or filled time.[3] Everything we have said also applies to the latter. For instance, if a sound continues for some time we cannot *perceive* its duration unless the end succeeds the beginning quickly enough to demarcate one unit of perception. The temporal limits of this perception are the same as in the case of empty intervals: 1.5 to 2 seconds. If a sound continues longer than this, there is no organized succession and in the end we reach the point where we perceive no change. The noise of a stream has no more perceived duration than the light of day.

---

[3] This terminology applies only to the physical description of stimuli. From the point of view of actual perception, the expression "empty" duration has no sense, any more than "empty" extension would have. We shall continue to use the expression, however, for the sake of convenience.

Duration is therefore only one of the characteristics of the organization of succession, and this is how we must interpret the statement made by Bourdon (1907) and Piéron (*La sensation guide de vie*, 3rd ed., 1955, p. 52), according to whom all sensations are of a temporal nature. This is simply because, as a general rule, every sensation is part of a succession.

The analysis we have just attempted may seem to be nothing more than the defense of a school of thought, but it is easy enough to show that authors whose opinions seem to differ widely from ours have started out from an analysis of the facts very similar to that which we have just made, only they have interpreted it differently. It would be simple to show that the discrepancies result from the fact that they introduced theories into their observations which are now outdated or merely that for them the psychological analysis was only a starting point.

In one way or another all these authors have recognized the importance of rhythm, or organization, in our perception of succession and in the interpretation of our "perception of time." They do not agree, however, on the central problem, that of the perception of duration. Duration of things? Duration of the ego? Duration as a combination of sensations or duration as a construction of our mind? Bergson opposes the material world characterized by plurality and externality to the spiritual world where we can apprehend pure duration, "what the succession of our states of consciousness becomes when our ego drifts through life and refrains from drawing a distinction between the present state and previous states." (Bergson, *Essai sur les données immédiates de la conscience*, 19th ed., 1920, p. 76.) But if we disregard these metaphysical elaborations, we realize that Bergson started out from psychological analyses which, long before the school of Gestalt psychology, acknowledged the importance of the organization of our sensations. "It is possible to conceive succession without distinction, as a mutual penetration, a solidarity, an intimate organization of elements, of which each one is representative of the whole and is only distinguished or isolated from it for a mind capable of abstraction." (Bergson, *ibid.,* p. 77.) He speaks several times of rhythm, characterizing it by the *quality* of a quantity where each stimulus becomes organized with its predecessors (*ibid.,* p. 80).

Wundt also admits, in a completely different context, that sensations of time are linked with rhythms and he thinks first of the rhythm of walking, to which vocal and auditory rhythms are subsequently associated. He agreed that we are able, through apperception, to apprehend a certain number of successive sensations at the same time; but being an associationist, he had to fill the intervals between exteroceptive sensations by other, interoceptive sensations. He imagined that these could come from the ears or even from feelings of tension and relaxation which would endow every sensation with temporal signs through which we could order them in time (Wundt, *Éléments de psychologie physiologique,* 1886). Münsterberg thought that these intervals could be filled by muscular sensations of attention and Schumann by a certain degree of expectancy (according to Bourdon, 1907). Obviously these authors postulate these sensations or feelings rather than prove them. If we have only basic sensations, there must be a bridge to link them: attention, feelings, or even other sensations of a continuous nature, such as sensations of tension or muscular sensations. In fact these authors are basing logical argument on the abstract idea of sensations which are apprehended at one point for one instant.

Mach also started out from rhythm to show the existence of a *time sense.* All the German authors of the nineteenth century used this expression. Some of them, such as Czermak (1857, quoted by Nichols, 1890), believed, in accordance with a post-Kantian trend of thought, that time, like space, is apprehended by a general sense distinct from the five special senses. Others used this expression of a time sense as a convenient formula for expressing our capacity for adaptation to time. Only Mach thought that a real sense of time analogous to the other five senses might exist. When we recognize the same rhythm in two different melodies, we have, according to him, perceived a scheme of durations independently of their sensory supports. This is possible only if we have perceived these durations in themselves, that is if we have a sense of time. Mach obviously understood that it was not enough to speak of a sense but that it was also necessary to define the receptors. In this field, however, he breaks hardly any new ground and his solutions are not very different from those of other authors such as Wundt. He thinks that there might be some kind of accommodation mechanism in the ear as there is in the eye and this is the sense organ for time. This organ would depend on stimuli and provide information concerning the temporal distance and position of the stimuli, just as visual accommodation does for distance and perspective. It would also depend on attention, for this effort would give rise to sensations of fatigue of the organ and these sensations would be a guide to duration (Mach, 1865, quoted by Bouvier, *La pénsee d'E. Mach*, 1923).

Janet was quite right when he called this "philosophical reasoning," for it consists of deductions based on not a single fact. Moreover, the problem of the sense organ does not stand alone. There is also that of the specific stimulus to be considered. If we speak of a sensation of duration "we assume that things exist outside of us just as we conceive them." (Janet, *L'évolution de la mémoire et de la notion du temps*, 1928, p. 47.)[4]

The fact that duration is a characteristic of the organization of succession may also be seen from an analysis of the part played by our different senses in the perception of change. Although we may perceive change with all our senses, the resulting perceptions of duration are not so homogeneous. Just as kinesthetic space and visual space are distinct, since they relate to the organization of different reactions, so the duration of a visual sensation and an auditory sensation cannot be directly compared. Despite this we can still speculate as to whether there is not, among the sense

[4] But Janet went too far. He was perfectly right in saying that our feelings of duration are reactions to the nature of our actions, but this very analysis prevented him from seeing that certain actions are in themselves direct adaptations to time just as there are adaptations to space. To apprehend succession as a unity, to be capable of synchronizing our movements to periodic stimuli as a dancer does, these are examples of such adaptation. The perception of change is not limited to that of pure multiplicity. It is organization, and the actions it results in are the basis of our adaptation to change. Reactions to the duration of our actions and attempts at conceptualization of our experiences of time come later.

organs, one sense which predominates when it comes to the perception of change. We should note first that changes are not perceived with equal frequency by all the senses: "We attribute duration to a sound because we always expect it to stop soon, but this is not so easy in the case of a color, because we are not so used to this changing." So said Herbart (quoted by Sivadjian, *Le temps*, 1938, p. 223).

We can consider the problem from another angle rather than from that of more or less rapid or frequent changes of the stimuli, by observing that our sense organs are adapted in very different ways to the perception of change. For this perception to be faithful, the temporal characteristics of our excitations must be very similar to those of the stimuli to which they correspond, that is to say the receptors must have little or no inertia. Let us take first the case of the organs of smell and taste, which have a high degree of inertia. The corresponding sensations are of indeterminate duration because the beginning and end of the stimuli are unclear. If several stimuli follow fairly rapidly on each other, they blend together without showing the temporal organization which implies discontinuity. This brings us back to the case of continuous change. The retinal receptors also have considerable inertia; sensations take a long time to become established and a long time to disappear. If successive stimuli follow rapidly on each other, they merge (cf. the projection of a cinefilm). If the frequency is slightly lower, it results in flicker. This is why although, theoretically, visual rhythms are possible as well as auditory rhythms, the former are of no use because the distinctness and clarity of the stimuli are too uncertain (Fraisse, 1948*b*). Moreover a rapid change of visual stimuli can be distressing. On the other hand the receptors of hearing and touch have practically no inertia. But touch can only give us information concerning changes which take place in contact with the body. The field is therefore limited (vibrations) but this property of touch has been used to teach deaf people (teletactor).[5]

Thus hearing is the main organ through which we perceive

[5](Translator's note.) An instrument which changes sound vibrations into touch vibrations.

change: it is considered as the "time sense" just as sight is that of space, this idea being based on a more or less explicit functional analysis. "Hearing only locates stimuli very vaguely in space, but it locates them with admirable precision in time. It is par excellence the sense which appreciates time, succession, rhythm and tempo" (Guyau, *op. cit.*, pp. 74-75).

As in the case of space, however, assimilation probably takes place between the information from the different senses, owing to the preponderance of one sense. We know that sight plays the leading role—except for the congenitally blind—in the case of space. And which is it for duration? Bourdon, who tried to analyze this, thought that, for himself, it was not sensations but "vocal imagery" which was predominant and acted as a sort of scale against which to measure all other durations. He recognized the fact that individual differences could exist, "that is to say that for some people the representation of duration may perhaps be more of a tactile nature, in others auditory and in yet others visual" (Bourdon, 1907, p. 477).

Seen from this point of view, the question is difficult to settle. One fact at least is certain; the sounds of speech provide us with a simple means of ordering our different, successive sensations. Through the apprenticeship of language we have acquired an amazing mastery over the organs of phonation, which, of course, lend themselves to this. We thus have a means of accompanying any series of stimuli by a succession of sounds which we *produce*. In this way it is easy for us to keep check of the order of stimuli and the duration of "filled" or "empty" intervals even if the sensations are not spontaneously organized. We have seen (pp. 76-77) that subjects who were asked to compare two durations defined by the sensory sequence sound-light-sound tended to synchronize with this a sequence of vocal sounds, such as "boom-boom-boom," thus recreating a unit of perception.

This predominant role played by vocal utterance should not surprise us; we understand only what we have recreated. It is all the more valuable because, owing to the successive nature of the temporal organization of perceptions, these never have the fullness of spatial forms which we can look at again and again.

III *The Perceived Present*

What we perceive, in time and in space, is an organization of
stimuli. This organization may be diffuse, and give us perception
only of an indefinite extension, such as when we look at a landscape without fixing our attention on a particular object, or perception of vague continuity, such as when we drift through life,
as Bergson puts it, without taking an interest in any particular
event. On the other hand, as soon as we fix our attention, organization comes into evidence, individual objects are distinguished,
successive structures are isolated and become figures against a
ground which remains indistinct. This organization implies unification, the demarcation of a group of stimuli which make one
whole, according to the laws established by Wertheimer for space
and the law of continuity for time (Koffka, *Principles of Gestalt
psychology*, 1935, p. 437). Unity is determined by the configuration of the stimuli but it is closely linked with the unity of the
actual act of perception which brings about the integration of
all the information received by the senses. We perceive succession only because, within certain limits, a "unified mental act" is
possible. The result of this unity of perceived succession—the
"tick-tock" of our clock—is the existence of a *perceived present*
which is not merely the passing of what was coming to be into
that which exists no more.

Broadly speaking, the present is that which is contemporaneous
with my activity. Obviously the changes to which it corresponds
are determined by the scale on which I see them. The present is
the century in which I live as much as the hour now passing. I
can actually make an arbitrary division of the changes in relation
to which I stand, by considering the past to date from a given
moment; I can, for instance, contrast past centuries with the
present century. This is the meaning of Janet's words: "The duration of the present is the duration of a story." (Janet, *op cit.*, p.
315.) There also exists, however, a *perceived* present which can
last only for the duration of the organization which we perceive
as one unit. My present is one "tick-tock" of the clock, the three
beats of the rhythm of a waltz, the idea suggested to me, the
chirp of a bird flying by. . . . All the rest is already past or still

belongs to the future. There is order in this present, there are intervals between its constituent elements, but there is also a form of simultaneity resulting from the very unity of my act of perception. Thus the perceived present is not the paradox which logical analysis would make it seem by splitting time into atoms and reducing the present to the simple passage of time without psychological reality. Even to perceive this passage of time requires an act of apprehension which has an appreciable duration.

All psychologists recognize the existence of this present, although they have called it by different names: the "specious present" (Clay, *The alternative*, 1882, quoted by James, *Principles of psychology*, 1891, vol. I, p. 609), the "sensible present" (James, *ibid.*, p. 608), the "psychic present" (Stern, 1897), "mental present" (Piéron, 1923, p. 9), the "actually present" (Koffka, *op. cit.*, p. 433). We prefer to call it the psychological or perceived present.

They have also given countless examples and comparisons to make their point. According to Wundt, just as our glance can cover only a certain extension of space, which can be varied according to how we direct our attention (I can look at the whole page of a book or simply at one letter, if I happen to be interested in the type of print) so in time "the field of vision of our consciousness" permits "the apperception of a series of successive, sensory stimulations." (Wundt, *op. cit.*, vol. 2, pp. 240-241.) James enlarges on his image of the flow of consciousness: "The only fact of our immediate experience is what has been called the 'specious' present, a sort of saddle-back of time with a certain length of its own, on which we sit perched, and from which we look in two directions into time. The unit of composition of our perception of time is a *duration*, with a bow and a stern as it were—a rearward- and a forward-looking end. It is only as parts of this *duration-block* that the relation of *succession* of one end to the other is perceived. We do not feel first one end and then the other after it, and from the perception of the succession infer an interval of time between, but we seem to feel the interval of time as a whole, with its two ends embedded in it. . . . The moment we pass beyond a very few seconds our consciousness of duration ceases to be an immediate perception and becomes a construction more or less symbolic." James, *Psychology, briefer course,* p. 280.) Piéron uses this image of time flowing like a stream: "There exists a *durable present* . . . in which we apprehend a succession of diverse facts in a single mental process which embraces, in the present, a certain interval of time, just as you can hold a certain amount of water in the hollow of your hand as it runs down from a spring; the water is renewed but the quantity is limited and can never increase." (1923, p. 8.)

The real problem is to know how to interpret this present. It is most often explained by a persistence of the elements which have just been perceived. This hypothesis is most clearly expressed by James: "Objects fade out of consciousness slowly. If the present thought is of ABCDEFG, the next one will be of BCDEFGH, and the one after that of CDEFGHI—the lingering of the past dropping successively away, and the incomings of the future making up the loss." (*Principles of psychology,* 1891, vol. I, 606.)[6] Wundt expressed the same idea: "With each new apperception, previous stimulations gradually retreat into the obscure depths of our internal field of vision and finally vanish altogether." (Wundt, *op. cit.,* p. 241.)

Although this interpretation underlies the thought of many authors, it does not concur with the facts we observe. The perception of aperiodic changes may delude us, but that of periodic changes does not permit us to explain the present by the varying degree of persistence of memory traces. When I perceive the "tick-tock" of the clock, I do not perceive first "tick-tock," then immediately "tock-tick," and so forth. What would become of a waltz rhythm if I perceived first a strong beat followed by two weak ones, then two weak beats followed by a strong beat, and finally one unit of weak beat—strong beat—weak beat!

Similarly, in the perception of speech, my present always consists of a whole clause, not the end of one clause followed by a bit of the next clause with a continuous glide that would make the whole sentence unintelligible. If we make an objective inventory at any given moment of the contents of the present, we will see that it is not composed exactly of the last elements

---

[6] This idea was later discarded by James in his *Psychology, briefer course.* We quote it only because it expresses this particular idea very neatly.

presented. If we read a series of ten letters to our subjects, they can grasp only six or seven—we shall return to this—but the six or seven they remember are not the last six or seven read. At the moment when we finish reading the series, the present of the subject consists of two or three groups of two or three letters from various positions in the series. "It would seem that the subject perceives successively several groups of elements, in the same way as we read letters in a text, that is not by sliding our glance along the lines but by discontinuous movements with stops from time to time, during which perception takes place. When the series does not contain more elements than we can grasp (six to eight), these discontinuities do not lead to mistakes any more than in reading. When the series is too long, however, there are gaps between the groups, just as when we read fast and the length of movement makes us skip words." (Fraisse, 1944-45.)

Confusion results from the fact that it is difficult to make an absolute distinction between memory and perception. "Memories are born in the very midst of perception," wrote Delacroix (1936, p. 327). This is shown too by psychological terminology, for the same phenomenon is called *immediate memory* or *perceptual span*, according to whether we are considering the possibility of immediate, integral reproduction or the perceptual aspect.

But immediate memory does not imply, as memory does, the existence of a past as a unit in relation to a present. "It is certain on the one hand that the different aspects of the present are not all on the same plane, as otherwise the present would seem static, and on the other hand that the present does not contain a unique element which has the nature of the present, all the rest being pure memory." (Delacroix, 1936, p. 313.)

In the case of spatial perception, it is obvious at once that immediate memory and the perceptual span are the same thing, for we check the extension of the perception of a subject by the number of elements he can name immediately after apprehension. The same is true for the perception of succession. If we can reproduce several elements after having heard them, this is not because we have memorized them but because we have a perceptual span which can cover several simultaneous or successive elements. We can grasp the same number of elements in both cases, and this confirms the truly perceptual aspect of this apprehension. Of course memory has something to do with it, but only indirectly. When I listen to a speech, I perceive the clause being pronounced by the speaker, but I interpret it in accordance with all the preceding sentences which I no longer perceive and of which I have only retained a general idea. When I listen to music, I perceive again and again a short, rhythmic structure, but this is integrated with a melodic whole to which it owes its affective resonance.

The fact that we are thus able to perceive several successive elements does not mean, however, that we can regard the perceived present as corresponding to a fixed capacity or to a standard duration of appreciation. From this point of view, the metaphor suggested by Piéron of a certain amount of water held in the palm of the hand, and the example given by James of the constant apprehension of the same number of letters may be misleading. The duration of the perceived present, like the richness of its contents, depends on the possibilities for the organization of successive elements into one unit. It is primarily determined by the direction of our attention. Here Wundt's comparison comes into its own. Within the field of vision (maximum field) there is one point of vision which can be larger or smaller depending on what we wish to perceive. On a page of print it may be one letter, one word, or even one expression. Bergson has the same idea: "This attention is something which can be lengthened or shortened like the distance between the two points of a pair of compasses." (*La pensée et le mouvant*, p. 191.) But this pair of compasses cannot go on being opened indefinitely; here we do not agree with Bergson who seems to think that the extent of our present depends entirely on our will: "My present at this moment is the sentence which I am engaged in expressing," he says, but he adds: "This is so because it is my will to limit the field of my attention to one sentence." (*Ibid.*, p. 191.)

In actual fact this field, which can be reduced to a single and more or less instantaneous sensation, also has an upper limit. This depends on several factors which may be summarized as follows: (1) the temporal interval between stimuli; (2) the number of stimuli; and (3) their organization. These three factors are actually interdependent but we must consider them separately in order to determine their exact nature.

1. The Interval Between Stimuli

Let us take a case where there are only two stimuli. If the interval between the two is too long, one is past when the other is present. This would be true of a clock whose "tock" followed several seconds after its "tick." When is this limit reached? One means of determining this is to slow down the succession of sounds in a rhythmic structure until the latter disappears and there is nothing but a succession of independent sounds. The rhythm is found to disappear when the interval between sounds is about 2 seconds (Fraisse, *Les structures rythmiques,* 1956, pp. 13, 41). This value represents a limit for any successive organization of two stimuli. Within the limit there is an optimum interval of succession which Wundt (*op. cit.,* vol. 2, p. 242) estimated at between 0.3 and 0.5 seconds. It will be noticed in music that the notes on which the organization of the melodic theme is based usually vary in duration, according to the composer and the piece, from 0.15 to 0.90 seconds (Fraisse, *op. cit.,* p. 118). When reading out loud, we pronounce between three and six sounds per second, which corresponds to an interval of between 0.15 to 0.35 seconds (see Chapter 5).

2. The Number of Stimuli

Taking the example of the chimes of a clock, we have already shown that when we hear 3:00 or 4:00 strike we identify the hour immediately without having to count. The perception of the whole permits its immediate interpretation as a symbol or the exact reproduction of a series of strokes, without having counted them. Thus young children who have not yet learned to count can reproduce a series of five or six strokes without making mistakes (P. Fraisse and R. Fraisse, 1937). On the other hand, when the clock strikes midnight we have to count; when the last notes of the twelve are ringing, the first ones no longer belong to our present.

If this is so, how many sounds can we perceive in one unit of time? At this point we shall discuss only identical sounds, but we shall see later that the variety and significance of the elements must also be taken into account. This question cannot be considered independently of the interval between the sounds or, if it is preferred, of their rate of succession. The length of the series which can be apprehended does in fact decrease as the intervals grow and the organization of the elements becomes more difficult. We found in our experiments that the average number of sounds apprehended, checked by immediate reproduction in the form of tapping, varied as follows, in the mean found for 10 subjects. (P. Fraisse and R. Fraisse, 1937.)[7]

Interval between sounds	.17 sec.	.37 sec.	.63 sec.	1.2 sec.	1.8 sec.
Number of sounds apprehended	5.7	5.7	5.4	4.0	3.3

This table confirms the fact that the intervals between .15 and .70 seconds are the most favorable for perception; it also shows that the unit perceived depends more on the number of elements than on the total duration of the series. The total duration of the series apprehended, counted from the first sound to the last, is 0.8 seconds for an interval of 0.17 seconds and 4.2 seconds for an interval of 1.8 seconds. We can therefore say that the duration of the perceived present varied more than the number of elements perceived; this shows clearly that the present does not simply correspond to a temporal field which is independent of its contents. This number of five or six elements, which marks the limit of our capacity for perception of succession, is also found when we apprehend stimuli of other kinds. For example, Pintner (1915) showed that, in the Knox cube test, the standard capacity for adults is the reproduction of six movements executed with no logical sequence. Similarly, if adults are asked to point to lamps, arranged in a circle, in the order in which they were lighted,

[7] The verbal or motor method of reproduction has been criticized as a means of checking the extension of the perceived present. Obviously reproduction cannot take place until after perception, when this present has passed away. But we are in exactly the same situation when we study the perception of space. In both cases we check on what is perceived through a reaction set off by the perception; this does not really imply mnemonic fixation. We have in fact shown that the processes underlying this immediate reproduction and the act of remembering only a few seconds later are largely independent (Fraisse and Florès, 1956).

they are successful within a limit of five lamps (Gundlach, Roths-child, and Young, 1927).

The number of these elements is a general characteristic of the span of our perception, for we can also perceive six or seven distinct elements in space, such as points of light (P. Fraisse and R. Fraisse, 1937). This span has guided men intuitively in their invention of sound signals. In the Morse code, for example, no signal contains more than five elements, and combinations of one to six dots are used in the Braille alphabet.

It is interesting to note that this perceptual span appears to correspond to a fairly general biological capacity which, in this basic form, is more or less independent of the level of intelligence. We have already shown that children of 4, 5, and 6 years of age who cannot count have a perceptual span comparable to that of adults. The same span has even been found in birds. It is possible to train pigeons, magpies, crows, parrots, or jackdaws to seek out from several cups the one which contains food, according to the number of dots on its lid. For instance, the birds will single out a cup with 5 dots on the lid from others with 3, 4, 6, or 7 dots. Naturally any possible influence from the configuration of the dots is eliminated by varying their arrangement. This experiment is only successful if the number of dots constituting the positive signal does not exceed 6 or 7. Results of the same sort are obtained if a technique is used which involves the successive perception of a number of elements. For instance, magpies can be taught to lift the lids of a line of cups containing one or more seeds until they have found a given number. This training is possible up to 7 seeds (Sauter, 1952). Arndt (1939-40) presented the rewards successively on a revolving disk and established by this method that the perceptual span is relatively independent of the interval between the stimuli, but these intervals must be shorter the greater the number of stimuli to be apprehended. This coincides exactly with the results found for children and adults when counting is excluded.

3. Organization of the Stimuli

It is a known fact that far more elements may be perceived in space if they form a spatial configuration or a unit of significance.

The same is true for time. If identical sounds are grouped, for instance in twos, threes, fours, or fives, we can perceive four or five groups of these sounds without counting. This means that, in the most favorable case, a total of 20 to 25 sounds can be perceived (P. Fraisse and R. Fraisse, 1937). Dietze (1885) made his subjects group the sounds subjectively and found a maximum capacity of 24. This result can only be obtained, however, if the speed of succession facilitates the grouping, according to Wertheimer's law of proximity. The results given above were obtained with an interval of 0.18 seconds between sounds and 0.36 seconds between groups. Therefore, the maximum duration of the apprehended series did not exceed 5 seconds.

The fact that each element has a different significance is in itself enough to permit a slight increase in our perceptual span. We perceive 7 to 8 letters and 7 to 9 figures if these do not form known words or numbers, but there is a genetic development in the span of apprehension of such stimuli. According to the norms of the Terman-Merrill test, at 3 years of age a child should reproduce a series of 3 figures; at 7, 5 figures; at 10, 6 figures. The older the child, the better identified and the more meaningful the elements; thus a certain link is apprehended between the stimuli and this facilitates their perception.

If the organization of the elements also gives a unity of meaning to the whole, apprehension is obviously improved; thus, still according to the Terman-Merrill test, an average adult should be able to apprehend and repeat without error a sentence of 20 to 25 syllables.

As our perception of succession is dependent on the possibilities of organization, everything which facilitates this—the attitude of the subject, grouping by proximity, structure, meaning—increases the richness of what constitutes our present. As we have seen, however, this present is doubly limited, by the interval between the elements and by their number. In view of those factors we have discussed, the present is limited for all practical purposes to a duration of about 5 seconds. This is the time necessary to pronounce a sentence of 20 to 25 syllables; the longest lines of poetry and the longest bars of music could scarcely exceed 5 seconds (Bonaventura, *Il problema psicologico del tempo,*

1929, pp. 33-34). In certain special cases it may be possible to attain a slightly longer present, but far more often our present consists of only 2 or 3 seconds.

As regards the physiological mechanism of this apprehension of plurality in one act of perception, we are reduced to the most tenuous hypotheses.

The psychological present has often been compared with the fluctuations of attention.[8] Fluctuations in efficiency are in fact very frequently observed; it seems that we are not able to maintain a constant level of activity. This is particularly noticeable in the case of perception. A threshold stimulus is perceived for a few seconds, then seems to disappear and reappear (the watch test, Masson's disk). In reversible figures, where several forms or several aspects of the same form can be perceived (e.g., the diagram of a cube showing all its angles of intersection), an alternation between the possible figures arises in perception as if there were saturation of one perception and the substitution of another figure. ". . . as if concurrent perceptual orientations could take it in turns to predominate." (Piéron, *L'attention*, 1934, p. 33.) These alternations can obviously be related to the general phenomenon of our perception of succession. It seems that after comparatively continuous perception, there is a break and then a new present begins. The periodicity of these alternations is also comparable; extreme periods varying from 5 to 10 seconds are found in different types of perceptual fluctuations (Piéron, *ibid.*, pp. 28-33). These durations are also approximately those of the perceived present under the most favorable conditions.

The periodicity of these fluctuations in attention depends a great deal on the individual, his attitude, and the conditions of perception. The duration of the perceived present seems to be influenced by the same factors. But all these comparisons explain nothing at present. We should have to prove that fluctuations in perception and the duration of the psychological present are effects of one and the same cause. This would only be possible if we could explain the nature of these fluctuations by discovering

[8] In particular by Bonaventura, *Il problema psicologico del tempo*, 1929.

the physiological processes to which they correspond. If this were done, we could believe in the existence of cycles of activity which facilitate the organization of succession.

Other authors have suggested more explanatory hypotheses. According to Piéron, "it may be that the extension of this field is related to the maximum time during which a brief cortical response can continue to promote the same process of association, which goes on like an echo even when other reactions are taking place." Piéron, 1923, p. 11.) Boring (1936) also considers the possibility of a physiological continuity between excitations, but this does not explain the periodic breaks. Koffka evolved a more precise theory: "The organization of two taps into a pair depends, according to our argument, upon the dynamic interplay between an excited area and the trace of a former excitation." (*Op. cit.*, p. 441.) He assumes that successive stimuli affect different points of the brain and the temporal interval is thus converted into a spatial interval. A difference in potential appears between the trace of the first sound and the excitation corresponding to the second, and a short circuit results which explains the temporal organization and the ordering of the sounds. The essential difference between a temporal and a spatial organization is that there is a difference in potential between the excited areas in the case of time which does not occur between simultaneous excitations in space.

What should we think of this hypothesis? It is natural to assume some kind of organization on the physiological level in order to account for perceptual organization. We know that the Gestalt hypotheses, which refer constantly to fields of force, do not concur with neurophysiological facts, but it is true that systems of neuronic interconnections may perhaps have a role equivalent to that of potential fields, as Hebb (*Organization of behavior*, 1949) tried to establish. Koffka's attempt to explain temporal organization by a theory of the spatial projection of successive excitations seems particularly weak. All his writings show that he was led by a desire to keep his system coherent, but the reason he gives for this cortical spatialization of succession is not convincing. According to him, if the second excitation were to occur at the same place as the first, which of necessity left a trace, this trace would be modified to such an extent by the second excitation that it would lose its identity, and the perception of a pair of sounds would be impossible. Of course if one stimulus follows very rapidly on another, the excitations and corresponding sensations do fuse more or less completely, but the problem of the organization of successive stimuli concerns temporal intervals where there is no fusion. We think it is more important to explain the permanence of the first process when the second takes place, which permits the organization of two successive stimuli.

We believe—for reasons which will be discussed in Chapter 5—that cortical excitation continues in a subliminal form beyond the time attributed to the sensation. The organization of excitations at one and the same point is possible without the sensations losing their individuality, if we take it that they are linked at a subliminal level and that each stimulus is identifiable through separate associative processes. This hypothesis, like those of Piéron, Boring, and Koffka, does not explain why there is a limit to this organization and why the break depends on the number of elements perceived and on the intervals between them. We should perhaps assume at this point the wave of activity which we referred to above, which could account for this limitation as well as for the fluctuations in the field of perception. But as yet neurophysiology gives us no help on this point.

Pathology throws more light on this whole question of the psychological present. To begin with, it reveals the fact that simple perception of the order of succession is a very basic capacity which is rarely upset by the most serious neuropsychiatric disorders, even when these involve temporal disorientation. The patient can still tell whether a sound precedes or follows a light (Fraisse, 1952b).

In many cases of mental disorder, however, the patient cannot perceive such long series of sounds as a normal adult. But there appears to be more than one possible cause for such failures and an analysis of them seems to confirm our interpretation of the psychological present. Some are simply due to the fact that as soon as the successive organization of stimuli lasts any length of time it requires an effort of attention or "presence" of which most neurotics are incapable. This point is stressed by every author, although explained in different ways: by a drop in psychological tension (Janet), a weakness in the nervous system (Pavlov) or a disorder of the conative component of personality (Eysenck). In other cases the cause seems to be a specific disturbance in the integration of succession; these are cases of cortical lesion. The disturbances are often difficult to define. There may be difficulty in perceiving series of sounds, i.e., rhythmic structures. "One of the commonest defects produced by a cortical injury is this lack of temporal definitions; a stimulus rhythmically repeated 'seems to be there all the time'." (Head, *Studies in neurology*, 1920, p. 754, quoted by Koffka, *op. cit.*, p. 438.) Van Woerkom considered disturbances in the perception of rhythm as one of the fundamental disorders of aphasia. For example, patients with aphasia cannot grasp the structure of iambics or trochees (quoted by Ombredane, *L'aphasie et l'élaboration de la pensée explicite*, 1951, pp. 243-255). From this Kleist inferred that this difficulty in perceiving temporal forms constituted a specific disorder (*Gehirn-Pathologie*, 1934).

Experiments have shown that patients with a damaged brain or a neurological disorder have difficulty in perceiving apparent motion, which is—as we shall see in Chapter 4—a form of integration of successive information (Werner and Thuma, 1942). In such cases the perception of each element seems to develop independently. In our opinion this independence may be explained to some extent by the fact that the process of preception takes longer. If the perception of succession implies the organization of distinct processes, an exaggerated duration of each perception would hinder or prevent temporal integration (rhythm, apparent motion).

Once again it is interesting to compare these cases with disturbances in spatial perception. Everyone who has carried out research on agnosia agrees that patients with this disorder fail to integrate perceptual information and this causes the dotted reproduction of forms or confusion and disorder of the elements. Disturbances in the perception of spatial and temporal forms are often found in the same patient (Teuber and Bender, 1949). This coincidence is easy to understand if we remember that there is an important temporal element in the perception of spatial forms. If a form is seen briefly it will appear to be either blurred or very lacking in detail. An appreciable time of vision is necessary to permit exploration and thus allow the form to appear in all its complexity.

Pathology shows us, therefore, that cortical lesions can prevent the organization of successive stimuli into temporal forms, but we cannot yet determine the exact location of these disturbances. It also appears that neuroses can reduce the efficiency of this organization of elements when they are complex and require an effort to be perceived correctly.

IV *Conclusion*

We think we have shown without any doubt that we can perceive the train or succession of changes provided that the interval separating them is not too long. They become grouped in units which form an almost static synthesis of change.

If we compare these forms with those we can perceive in space, we see that they are relatively simple; it is easy to see why, for we cannot go back in time to develop an analysis which would permit of a more complex construction. For this reason the arts limited by duration seek to enrich each moment: harmony combines with melody, various instruments add their different tones, dance or song accompanies music. Thus the simultaneity of excitations compensates for the poverty of perceptible successions.

But one unit of perception succeeds another. Between them there is a slight lapse, a pause that we do not even notice; in speech its presence is marked by punctuation. But discontinuity in perception is disguised by the continuity afforded by the emotional quality of events and their unity of significance. Each unit perceived takes its place in a stream in which the durability of our attitude and our memory are the factors which determine continuity. The part played by meaning is obvious in the case of language, whether poetry or prose; in music the discontinuity of the rhythms is scarcely noticed, for each unit is part of a musical flow which gives unity to the whole.

Thus through our psychological present we are masters of change in the world of stimuli. Through it we perceive units of change which in their turn are elements from which we construct the unity of our whole psychological life.

||||||||||||||||||||||||||||||||||||||||||||||||||||||||||||||

THE THRESHOLD OF TIME

THE EXISTENCE OF THE PSYCHOLOGICAL PRESENT IMPLIES THAT several successive events can be apprehended with relative simultaneity. In other words, where the temporal variable of change is, physically speaking, continuous, psychology shows a discontinuous integration of successive elements into a series of perceptions. There is nothing surprising in this. All perception consists of phenomenal information which corresponds to its stimuli in its quality and organization but is not a mere copy of physical reality. The psychology of perception consists in establishing these psychophysical correspondences and trying to explain them by finding the mechanisms of reception, transmission, and cortical projection.

The first question regarding any perception is that of its threshold. Under what conditions does time appear as a perceived fact? That is the problem we wish to discuss in this chapter; in the next we shall study the variations in our perception of duration according to the nature of the stimuli.

Two typical situations may arise:

1. The perceived change is continuous, and we also perceive continuity. In this case, if the physical stimulus is brief, we do not perceive duration but instantaneousness. What must the duration of the stimulus be before we pass from the instanta-

neous to the durable? In other words, what is the threshold of duration?

2. The stimuli are brief and repeated. In this case the following question arises: to what physical interval does the perception of succession correspond? In other words at what interval do two occurrences cease to appear fused or simultaneous?

Instantaneousness and simultaneity are the two limits at which perception of time ceases. Inversely, if we study the conditions under which perception of the instantaneous and the simultaneous cease, we shall see the beginnings of the perception of time.

I *From the Instantaneous to the Durable*

"Thus when we feel one instant on its own, instead of feeling it as before and after in movement, or as the end of before and the beginning of after, it seems that no time has passed because no movement has occurred." This definition of Aristotle's still stands (*Physics*, book IV). A brief stimulus can be perceived without appearing to be durable. Here we are at the limit, which Piéron calls a "point of time" by analogy with space (*La sensation guide de vie*, 3rd ed., 1955, p. 401).

All "non-durable" sensations are theoretically identical in respect of time. But in practice, when the physical duration of stimuli decreases, the apparent intensity of the corresponding sensations is also lessened. This intensity is in fact proportional to the quantity of energy received by the sensory receptors; that is, to the product of physical intensity and the duration of the excitation. The difference in intensity, therefore, permits differentiation between sensations and thus prevents us from confusing them although they are instantaneous.

On the whole, however, instantaneous sensations are distinct from durable sensations, and it is possible to determine the limit between the instantaneous and the durable according to the duration of the stimulus. Durup and Fessard (1940) found that the threshold of duration was 0.124 second for a stimulus of light of 1 millicandlepower/cm.^{2} and 0.113 second for a luminosity of

100 millicandlepower/cm.[2] The same authors found thresholds varying from .01 to .05 second for a sound stimulus of 500 cycles per second and average intensity. Before them Bourdon (1907) had found a threshold of .01 to .02 second for the same conditions.

According to the preliminary experiments carried out by Durup and Fessard, there is also a threshold of a few hundredths of a second for tactile sensations produced by a vibratory stimulus. Following these authors, we should note that the limits for the perception of instantaneousness depend on the duration of the whole process of excitation. The maximum value for the *point of time* expressed in terms of the duration of the stimulus is far longer in the case of vision than it is for hearing or touch. But we know that the photochemical processes of excitation of the retinal receptors have far more inertia than those of the auditory or tactile receptors which are of the mechanical type. The time necessary for activating the process of excitation does not count from the point of view of perception. The important factor is definitely the duration of the cortical excitation, which is to determine the temporal characteristics of the perception. Piéron thinks that it is the central processes which cause "a minimal extension of this point around a hundredth of a second." (*Op. cit.*, p. 403.)

This *point of time* has also been considered as a psychological unit or atom of time. The question of the existence of a psychological unit of time has often been raised, but with very varied interpretations. Some authors, including Piéron (1923, 1945), have considered it from a psychophysical point of view: what is the simple, or indivisible, element of time? As we have seen, the unit of time varies according to the nature of the sensation. A number of authors take a psychological unit of time to mean the minimum duration of the simplest possible mental operation. Richet (1898), who was the first to suggest this, noted that we cannot pronounce more than 11 syllables or vowels in one second, which seems to indicate that there is a limit to the frequency of central excitations. He tried to show by a great many examples that this limit of one tenth of a second is found in many psychic manifestations. According to him it corresponds to a basic nervous *oscillation* whose duration is determined by that of the refractory period.

What is in this case called a psychological unit is in actual fact the minimum duration of the physiological process corresponding to a simple action, without reference to the apparent duration of this action on the level of perception. Even very recently, Stroud (1956) tried to show that psychological time, or the time of psychological activity, could only be divided into a *finite* number of moments, whereas psychic time could be split up into an infinity of instants. His aim is the same as Richet's but he approaches it from another angle and carries out a variety of experiments in an attempt to measure the duration of a psychological *instant* or interval of integration which, according to him, should also last about one tenth of a second. Let us take two of his examples. If a subject is presented with a series of very brief auditory or visual stimuli (stimuli lasting 11 milliseconds with a break of 22 milliseconds, i.e., following at intervals of 33 milliseconds), the number of stimuli perceived is less than the objective number. This may be compared with a cine-camera whose shutter will not open more than so many times per second, with the result that only part of what is happening can be recorded (White and Cheatham, quoted by Stroud). If, while the subject is listening to a list of words, the flow of sound is interrupted electronically at a certain frequency (or covered by a white noise) the effect of these interruptions is found to vary considerably according to their rate. If this is slow and the duration of the break is equal to the duration of the sounds, only 50 percent of the words are perceived. If the rhythm is very rapid nothing is lost and there is 100 percent success. Between these two extremes we find that success approaches its maximum as soon as the number of interruptions reaches 10 per second, as if practically no useful information were lost at this frequency (Miller and Licklider, 1950).

Other authors have taken the instant, or irreducible instantaneous unit, to be the interval which just permits us to distinguish between repeated stimuli, so that there is no fusion. But Piéron, who quotes this as the standpoint of Uexküll's pupils, points out that this frequency of fusion depends mainly on the receptors and therefore cannot be used for the accurate measurement of a central process (Piéron, *Psychologie zoologique,* 1941, p. 102). On the other hand we may be able to find a significant instant in the passage from discontinuity of sensations to flicker, for this takes place for all the human senses when the interval between successive stimuli reaches 0.05 second (Brecher, 1937).

The object of such research is to find a unit for the physiological processes of integration of succession; the facts on which it is based lend likelihood to the existence of such a unit. But the problem cannot be completely solved without progress in the field of the neurophysiology of the higher centers. If we knew the nature of the physiological unit of time, we could probably define the conditions under which we can perceive the instantaneous. But can we go further and speak of a psychological unit of time? The word *unit* has two meanings:[1] first it is that which is indivisible; in this sense the instantaneous may be called a unit of time. Second, it is a multiplicity of parts constituting one whole. Can both these meanings be applied legitimately to time? This is doubtless the idea behind Piéron's words (1945, p. 36): he speaks of durations composed of "unified pluralities of instants" and raises the question of the relationship between the psychological unit of time and the value of the scale interval in the comparison of perceived durations. Can there be quanta of perceived time? Psychophysical research has not given us the answers to these questions. And there is nothing to prove that perceived durations are compound.

In the first two chapters of this book we showed that there must be a physiological clock based on the property of the nervous centers to respond rhythmically to excitations, whether these are periodic or not. If we knew more about the mechanism of these rhythms and their basic frequencies, we should perhaps be able to solve the problems we have just discussed.

---

[1] (Translator's note.) The word *unité* in French covers both *unity* and *unit* in English.

II *From Simultaneity to Succession*

When there is a single stimulus, its duration is perceived if it lasts long enough, so that it no longer appears to be instantaneous. On the other hand, two brief stimuli permit perception of duration when they appear to be successive. The duration is in this case the interval between the two stimuli. If there appears to be no interval between the two stimuli, they are said to be simultaneous. No time elapses between them.

The following question now arises: under what conditions do we perceive simultaneity? And correlatively, what are the thresholds of the perception of succession?

1. Simultaneity

It is common sense to say that two events are simultaneous when they occur at the same moment of time. But as Poincaré observed very acutely (*La valeur de la science,* pp. 39-63), this is tantamount to claiming an infinite and omnipresent intelligence; for man—who from this point of view acts like any recording instrument—never has a direct knowledge of physical phenomena but only of the sensations produced by these phenomena. The order in which these physical phenomena take place does not determine the order of our sensations. Lightning, produced by the discharge of electricity from a cloud, and the sudden vibration of the air which it causes take place at the same time, but in our perception the thunder follows the lightning. On the other hand, two flashes of lightning may seem to be simultaneous whereas the flash which occurred nearer the observer actually took place after the more distant flash.

We are concerned here with psychological simultaneity when "events belong to the same mental present and cannot be ordered in time." (Piéron, *La sensation guide de vie,* p. 394.) Our main problem is to determine the relationship between this apparent simultaneity and the actual order of the physical phenomena as far as this can be established through other methods of recording than that of our own senses.

The most obvious reason for this discrepancy between the perceived order and the physical order of events is the difference in speed of transmission in the external world of vibratory phenomena, and, in particular, the difference in the speed of sound and light. It is the task of physicists to determine the duration which elapses between emission and reception; when great distances are involved this can be very difficult, as the theory of relativity has shown.

In addition to this discrepancy, we must also consider the differences in the duration of the processes of peripheral excitation and in the speed with which the sense organs transmit these to the cortical centers of perception. Here we meet the biological aspect of the problem. Two stimuli which reach the organism simultaneously are not necessarily perceived simultaneously. There are a number of reasons for this.

Some are physiological. To begin with, each type of receptor has its own latent period. In the most favorable circumstances, the irreducible latent period of vision is 0.04 second greater than that of hearing (Piéron, *ibid.*, p. 46). The latent period of one receptor may also vary according to the intensity of the stimulus. The greater the latter, the smaller the latent period; "the reducible portion of the latent period is inversely proportional to the intensity of the stimulus when this is brought to a certain strength either lower, equal to or higher than the unit." (Piéron, *ibid.*, p. 467.) The result, to give an example, is that if two small areas of light, fairly near each other, are lighted simultaneously but at different levels of intensity, they will not appear to be simultaneous; the more intensely lighted area will seem to move toward the less brilliant area. This apparent motion is perceived when two similar stimuli follow each other in fairly rapid succession; the phenomenon is characteristic of the borderline between simultaneity and succession.

Apart from the latent period of the peripheral organs, another factor which may lead to a discrepancy in the time of perception of two stimuli which are objectively simultaneous at the level of the receptors is the delay in the transmission of the nerve impulse from the periphery to the center; this is due to the duration of conduction in the nerves and crossing the synapses. Klemm (1925) showed that if two stimuli, one on the forehead and the other on the thigh, are to be perceived simultaneously, the excitation of the thigh must precede that of the forehead by between 20 and 35 milliseconds; this time corresponds more or less exactly to the time necessary for the nervous impulse to cover the difference in the length of the fibers connecting the thigh and the forehead to the cortex.

These facts prove the validity of previous theoretical reasoning. The perception of simultaneity depends on the simultaneity of cortical excitation.

At this level, however, there are other more specifically psychological factors to be considered, particularly that of the direction of attention. Basing his observation on Wundt and James, Titchener noted that "The stimulus for which we are predisposed requires less time than a like stimulus, for which we are unprepared, to produce its full conscious effect." (*Psychology of feeling and attention*, 1908, p. 251.) The result is that, of two stimuli which act on the organism under the same conditions, the one toward which we direct our attention will seem to precede the other. Bethe showed that if a Geissler tube is placed behind a row of small windows, the illumination will seem to emanate from the window at which we are looking (quoted by Fröbes, *Lehrbuch der experimentellen Psychologie*, 1935, I, p. 386). Similarly Piaget found that if children look at two lamps (at a distance of 1 meter and symmetrically placed in relation to the mesial plane of their bodies) which are lighted simultaneously, 80 percent of their errors are due to the fact that they think the light they were looking at was lighted before the other (*Le développement de la notion de temps chez l'enfant*, 1946, p. 120).

These experiments do not actually prove that the delay of one sensation in relation to another is due to *attention*, for in the case of vision the delay in the perception of the stimulus which is not being looked at may be due to a greater latent period of the peripheral receptors of the retina. It has been estimated in the most recent research work on this subject that the delay in a peripheral stimulus as compared with a foveal stimulus, with light adaptation, is 10 milliseconds at 10° and 20 milliseconds at 40° from the fovea (Sweet, 1953). But this factor is obviously not enough to explain similar results found when it cannot have any influence. Stone (1926) carried out a very precise experiment to measure the degree of influence of attention. He aimed to establish the limit of duration of the interval for which an auditory and a tactile stimulus seem simultaneous when the subject fixes his attention (a) on the sound, (b) on the touch. He obtained a value of approximately 50 milliseconds which shows the role played by attention. This same value was found by Rubin (1932): he simply changed his instructions to the subject—which obviously affect the latter's attitude—in order to determine which of two stimuli separated by 50 milliseconds would be the first to be perceived.

The full significance of the specific role played by attention is seen only if we remember that it is impossible to pay attention to two things at once. This fact is proved by countless experiments. For instance, if a subject is asked to identify a group of dots or lines shown through the tachistoscope and, simultaneously, a group of punctiform tactile stimuli, he will fail, whereas if the two groups are presented separately he will easily succeed. An interval of at least 0.2 second is necessary for the accomplishment of this double task and the optimum results are obtained for an interval of 0.6 second (Mager, 1925). Using more introspective methods, Feilgenhauer (1912) found that the interval between two stimuli must be between 0.2 and 0.4 second before the subject considers he has time to transfer his attention from the first to the second stimulus.

Since we are incapable of fixing our attention on two stimuli at once, we direct it toward one or the other, either spontaneously because it attracts our attention, or voluntarily. And the stimulus to which we turn our attention seems to precede the other. Thus attention modifies the duration of processes of perception either by accelerating one or by temporarily inhibiting the other; this gives rise to a temporal discrepancy, which makes a succession of two excitations which reach the brain simultaneously.

This point of view is confirmed by most of the work carried out on the perception of simultaneity. True perception of simultaneity cannot take place unless the stimuli can be integrated or unified so that we apprehend them *together* without having to divide our attention. Inversely, whenever this unification is difficult, the perception of simultaneity is very unstable.

Let us take these two points separately. If two or more stimuli form a figure with one unit of meaning, their simultaneity presents no problem. The near fusion of two notes in a musical chord permits the perception of perfect simultaneity. On the other hand it is difficult to decide on the simultaneity of a knock at the door and the clock striking the half hour, because there is no relationship between these two sounds. Unity may result from a condition outside the stimuli themselves. A brief flash of lightning which illuminated from the outside the two areas of Piaget's experiment would have resulted in one unit of perception and no doubt everyone would have agreed in this case that they saw the two stimuli simultaneously. Other experiments by Piaget have shown that the *perception* of the simultaneity of the cessation of two movements, for instance, is not stable in the case of young children unless the two movements are integrated in some way to form one unit of perception. "When two moving objects leave the same place and arrive at the same point at the same time, there is no difficulty in the way of perception of the simultaneity of their starting and stopping." (Piaget, *ibid.*, p. 105.) But if they stop simultaneously after travelling at different speeds and arriving at different points in space, the young child will fail because the cessation of each movement belongs to a different unit of perception.[2]

One method by which we can verify our perception of simultaneity is to include stimuli with no apparent connection in one unit of reaction. If we react to a sound by tapping with our left hand and to a flash of light by tapping with the right, for instance, the comparison is easier. We are capable of great accuracy in the appreciation of the synchronism of symmetrical movements which can be integrated into one motor pattern. We can make two simultaneous movements with great precision, the average discrepancy being no more than a few thousandths of a second, even if different and asymetrical limbs are involved (e.g., right hand and left foot), implying an initial discrepancy

[2] An older child can surmount these difficulties by reasoning and by comparing the points of departure and arrival and the speed of movement.

in the time of release of the motor stimuli. (Paillard, 1947–48.)

On the other hand it is very difficult to assess the simultaneity of two sensations which have nothing in common; this is true for stimulations involving the same sense, but even more so for heterogeneous stimulations. This problem was at the root of the first research carried out by experimental psychologists in the nineteenth century on the personal equation. It had been noticed by astronomers that they made mistakes in judging the moment at which a star passed a visual reference point (the cross hairs of the telescope) in relation to the successive beats marking the time. These errors tended to be systematic for any given person, hence the name "personal equation." Research on this problem was carried further by Wundt in his so-called complication experiment. His subjects were to judge the position of a pointer moving across a dial at the moment when a sound was heard. The results showed errors in this location of up to 0.1 second, the pointer usually being seen in a position which it actually reached after the sound was produced (Wundt, *Éléments de psychologie physiologique*, 1886, vol. II, p. 302). The same value for this error was found by several other authors, particularly Michotte (1912), who also showed that the direction of the error was determined by the conditions of perception. These have the effect of displacing the "point of attention" and Michotte states most emphatically that a stimulus is "apperceived" only at the instant when it is expected. The existence of this error shows how difficult it is to judge the simultaneity of two stimuli. Furthermore the difficulty is even greater in this sort of situation where a moving stimulus must be located in relation to the graduations on the dial.

A different kind of experiment, carried out by Guinzburg (1928), also illustrates the difficulty experienced in perceiving simultaneity and the kind of mistake that can be made. His subjects were presented with the task of saying whether a light and a sound were simultaneous or successive. First let it be said that two of his ten subjects proved incapable of giving coherent replies. When the stimuli were objectively simultaneous, Guinzburg obtained only 39.2 percent correct answers. The answer "Simultaneous" was given most frequently (45 percent) when the sound preceded the light by 30 milliseconds. Simultaneity was still perceived (in 8.3 percent of the cases) when the sound preceded the light by 120 milliseconds, or even when the light preceded the sound by the same interval (67 percent). This objective difficulty in perceiving simultaneity relates to the fact that there is no actual impression of simultaneity for stimuli pertaining to different sense modalities. "The simultaneity is inferred from the indistinctness and lack of a clearly imposed order which leaves room for a certain liberty of arbitrary arrangement." (Piéron, op. cit., p. 394.) Guinzburg asked his subjects to indicate the cases where they were sure of their judgment and obtained a smaller number of affirmatives for perception of simultaneity than for perception of succession.

The perception of true simultaneity, therefore, implies that the excitations can be organized into one pattern of perception or reaction. This organization permits at the same time the stable perception of simultaneity and a very fine threshold for the perception of succession. On the other hand lack of organization extends the zone in which there is nonperception of succession without the appearance of actual simultaneity. This zone obviously varies with the nature of the stimuli. It is 10 milliseconds for two auditory stimuli or two tactile stimuli but it is as much as 30 milliseconds between an auditory and a tactile stimulus and 80 milliseconds between an auditory and a visual stimulus. (Cohen, 1954.)

The necessity of this integration probably explains why people suffering from aphasia and more generally with any kind of cortical lesion, have difficulty in perceiving the simultaneity of a sound and a light in the most simple situations where no other mentally ill person fails (Fraisse, 1952b).

2. The Threshold of the Perception of Succession

At what interval of time do two stimuli cease to be confused or simultaneous? Piéron suggested the name of *temporal acuity* for "the capacity of discrimination in the dimension of time, just as spatial acuity is used for the capacity of discriminaton in the dimensions of space." (Piéron, op. cit., p. 394.) We must distinguish between three separate cases, as Piéron does (1923):

"1. The two stimuli are identical and act on the organism at the same point. . . .

"2. The two stimuli are identical but act on the organism at two different points, or are similar but not identical. . . .

"3. The two stimuli are very different (different sense modalities)."

In the first case, if the stimuli follow each other in rapid succession, they merge owing to the persistence of the sensation and we only perceive one more or less durable sensation.

If the temporal interval between the stimuli is slightly longer, we still perceive a continuous stimulation, but of varying intensity. Under these conditions a series of stimuli will give rise to the phenomena of flicker for vision, beats or crepitus for sound, and vibration for touch. These phenomena appear when there is a reduction in the intensity of the sensation between the stimuli at least equal to one scale interval. In this case we actually perceive a change rather than a true succession. The interval at which perception of continuity gives way to perception of a change in intensity varies with the nature of the receptors. The ear, for instance, is still able to discern interruptions in a white noise at a frequency of 1000 per second, that is when the interval is of 1 millisecond (Miller and Taylor, 1948). For touch, a vibration is still perceived with the tip of the index finger at more than 1000 stimuli per second (Piéron, op. cit., p. 68), and for vision, under the most favorable conditions, a flicker can still be observed at a frequency of 60 per second, i.e., at a temporal interval of 16 milliseconds (Mowbray and Gebhardt, 1954).

The threshold of true discontinuity is far higher and depends on the intensity of the stimuli; we can give only approximate values: 10 milliseconds for sound and touch, 100 milliseconds for sight (Piéron, op. cit., pp. 396-397).

When the stimuli act upon different points of the organism, we observe, between the perception of simultaneity and that of succession, an organization of successive excitations which gives rise to complex perceptions.

In the case of vision, the stimulation in rapid succession of two points on the retina nearly always results in the perception of apparent motion. We do not perceive the duality of elements but a single stimulus which moves from the place of appearance of the first to that of the second. When the phenomenon is known it is possible to infer succession from the motion, but we do not actually perceive succession. The temporal limits within which apparent motion is perceived vary considerably and depend on the intensity and distance of the stimuli, in accordance with Korte's laws. As we have already shown, it is even possible to produce apparent motion with two stimuli which are objectively simultaneous but are of different intensity, as a result of the greater delay in the case of the less intense sensation.

It is therefore difficult to fix values for the temporal intervals at which we pass from simultaneity to apparent motion and from apparent motion to succession. In the case of vision, however, we will recall that Wertheimer thought that the optimum interval for apparent movement was 60 milliseconds (the cinema has used a frequency of 18 and later 24 frames per second, i.e., intervals of 55 and 40 milliseconds), and that this disappears completely, giving place to the perception of succession, when the interval reaches 200 milliseconds. The threshold between simultaneity and apparent motion can have a very low value when circumstances are favorable. For instance, we can perceive apparent motion between two adjacent light stimuli acting on the fovea at intervals of 50 milliseconds (Sweet, 1953). Obviously this threshold is far higher under normal conditions, and it should be noted that it is even higher, as already shown, in the case of subjects with cortical lesions. Children with lesions have a threshold for apparent motion of more than 200 milliseconds, while for normal children of the same mental age it is only 75 milliseconds. These lesions help to isolate sensations and prevent their organization, thus extending the zone of apparent simultaneity and raising the threshold of temporal acuity (Werner and Thuma, 1942).

There is also some variation in the nature of perception in the zone between simultaneity and true succession in the case of touch. If two very close points on the forearm are stimulated successively (the distance between the two points being below the threshold for spatial discrimination of the two contacts), at intervals of less than one hundredth of a second, only one sensation is perceived, over a small area located nearer the first point to be stimulated (Klemm, 1925). When the intervals are slightly longer, apparent motion is perceived, as in vision (Benussi, 1917). Piéron gives an interval of around 10 milliseconds as the threshold for discrimination between stimuli applied at close or symmetrical points.

For sound the phenomena are more complex. Hisata (1934) observed the existence of apparent motion between two identical sounds which were brief, 20° apart, and separated by an interval of 20 to 60 milliseconds. A rapid succession of excitations applied to both ears results in lateralization of the sound. The threshold for this perception is very low: 0.07 milliseconds according to Aggazzotti (1911), 2 milliseconds according to several other authors (Piéron, *op. cit.*, p. 396).

To sum up, temporal acuity is greater when identical stimuli are applied successively to different receptors of the same sense organ and not to the same receptor cells. This is natural, for the inertia of the receptor no longer plays any role. The acuity also appears to be greater because it is based on the interpretation of perceptions of a change (apparent motion, impressions of one stimulated area, etc.) and not on the perception of a true succession; this has been shown very clearly by Sweet (1953). With peripheral vision at 20° from the fovea and with light adaptation, subjects can discern motion between two adjacent areas when there is an interval of 5 milliseconds, and they cannot see whether there is one light or two unless this interval reaches 39 milliseconds. If conditions are such that the two stimuli cannot be integrated to form a single phenomenon, temporal acuity is far poorer.

When the sensations pertain to different senses, the threshold for the perception of their succession is obviously higher since they cannot immediately be organized. The values given by various authors vary between 50 and 100 milliseconds (Piéron, *op. cit.*, p. 396 and Tinker, 1935).

It should be noted, however, that most of these values were found in the nineteenth century with apparatus deficient in accuracy and with rather elementary psychophysical methods. Moreover these measurements are very tricky for it is difficult to use sensations of the same intensity for the different senses. Another difficulty was also encountered. Exner, for instance, measured the smallest discernible time between a visual stimulus and an auditory stimulus and found a value of 16 milliseconds, but when the order of the stimuli was reversed he found 60 milliseconds (quoted by Wundt, *op. cit.*, vol. II, p. 295). This kind of variation led to a great deal of research. A few authors quickly realized that the explanation was that certain stimuli attract more attention than others and are therefore more rapidly perceived (Whipple, 1898). For instance, in the case we have just quoted, the auditory sensation must have attracted more attention than the visual sensation. The majority of authors, however, have generally found the opposite; light usually attracts more attention, all the more so since the source of sound is better hidden (Bald, Berrien, Price, and Sprague, 1942). We should not be surprised by this divergence of results, for the nature of each stimulus may modify the result and there can be appreciable differences between individuals.

Let us give one particular instance: that of the perception of succession between a motor response (press on a button) and an exteroceptive sensation, such as light. The threshold of succession is still much the same (50 milliseconds) as in the case of two exteroceptive stimuli (Biel and Warrick, 1949).

All through this chapter we have stressed the importance of the receptor processes in the discrimination of duration and succession. It would perhaps be useful to give a summary of the principal results in a table. It should be remembered that the figures given are not precise but approximate values.

If we consider the changes which take place on the human level, it is evident that the thresholds are very low and we are well equipped to perceive time in its two fundamental aspects of duration and succession. But, we perceive change better with our sense of touch and hearing than with sight, the latter being better adapted for spatial discrimination than for the apprehension of the temporal modification of stimuli.

	Threshold of Duration	Threshold of Succession	
		Stimulation Involving One Sense	Stimulation Involving Different Senses
Touch	10 to 20 millisec.	10 millisec.	
Hearing	10 to 20 millisec.	10 millisec.	50 to 100 millisec.
Sight	100 to 120 millisec.	100 millisec.	

ll

PERCEIVED DURATION

WE CAN PERCEIVE TIME WITHIN THE LIMITS OF THE PSYCHOLOGICAL present, but the modalities of this perception vary in quality and quantity with the physical nature of the changes perceived. We do not perceive duration independently of what endures. Starting out from this fact, we shall study in this chapter the modalities of our perception of time in its various aspects.[1]

I *The Quality of Durations and the Indifference Zone*

Between the lower limit, where we are first able to discern two distinct stimuli, and the upper limit, where one stimulus is already part of our past when the next occurs, there is perception of true succession and of an increasing interval between successive stimuli. If we were to leave it at this we should simply be transposing the physical measurements of the interval between stimuli on to the psychological plane. In actual fact, as this interval increases, the succession of stimuli gives rise to perceptions which are different in *quality*.

[1] Our readers will perhaps be surprised by the way we handle these problems, but such is the state of our psychological knowledge. Perception is one of the most simple kinds of behavior and has therefore been studied more than other more complex behaviors. It is therefore better understood.

1. The Quality of Intervals

At the moment when stimuli cease to be confused and appear successive and distinct, we cannot actually yet perceive an interval which is like a gap between them. At this point the sensations appear to be distinct, but *contiguous*. When the interval is a little longer, we perceive an *integrated pair* of stimuli (Schultze, 1908). Thus we do not spontaneously perceive a gap. We perceive two more or less closely linked stimuli. The interval is not perceived in itself, although it is discernible if we fix our attention on it. When the gap between stimuli reaches about 0.6 second we have spontaneous perception of an interval, but this is not dissociable from its limits. When the gap is greater than 1 second, the interval becomes the dominant feature and it requires an increasing effort to regard the two stimuli which delimit it as one unit defining the duration of the interval. Finally, when the gap reaches 1.8 to 2 seconds, the two stimuli cease to belong to the same present; we no longer perceive one interval of duration but only the distance between a past event and a present event.

We could give this same description in terms of the speed of succession. With the help of a metronome, Vierordt (*Der Zeitsinn*, 1868) found that the judgment "fast" corresponded on an average to an interval of 0.42 second, "neutral" to 0.64 second and "slow" to 1.07 second.[2] We can also consider the question from the point of view of the durations themselves instead of putting the accent on the intervals between stimuli. Katz (1906) distinguished three types of duration: the short, from 0.25 to 0.55 second, the "comfortable," between 0.60 and 0.65 second, and the long, above 0.65 second. Benussi was even more precise: very short durations extend from 0.09 to 0.23–0.25 second, short durations from 0.23–0.25 to 0.58–0.63 second, indifferent durations from 0.58–0.63 to 1.08–1.17 seconds, long durations from 1.08–1.17 to 2.07 seconds, and very long durations above this. If we

[2] We would point out that the values found by Vierordt are probably relative to the scale of intervals given by a metronome (see p. 120). Frischeisen-Köhler (1933b), also using a metronome, found that the *tempi* which were not judged to be either slow or fast corresponded to a zone of intervals from 0.55 to 0.83 second, according to the subject.

compare these different analyses without attempting to go into details which cannot correspond to qualitative judgments, we can distinguish between three zones:

1. Short intervals of less than about 0.50 second. For these durations we perceive the limits rather than the interval itself;
2. Indifferent intervals, that is neither short nor long, approximately from 0.50 to 1 second. In this case the limits and the interval form one unit;
3. Long intervals of more than about 1 second, where the perception of a gap is predominant and an effort is necessary to join the two limits in a single present.

We can apply the same analysis to filled durations, for instance to a continuous sound. The beginning and end of the stimulus correspond to the limits of empty time. For short spaces of time there appears to be no duration between the beginning and the end, which are the dominant phenomena; in the case of longer durations, the beginning and end cannot be dissociated from the duration, and for long spaces of time the duration predominates over the initial and final sensations.

We have based our analyses, like other authors, on auditory sensations. They can also be applied to tactile sensations but they gradually lose all significance for the sensations which develop slowly and thus do not facilitate the discrimination of stimuli or the perception of succession. In the case of smell, the question of a temporal interval between smells cannot even be considered.

The importance of these qualitative distinctions is due at least in part to the fact that there are different laws of perception for each of the three categories of time. As early as 1864 Höring, a pupil of Vierordt, found that, among intervals ranging from 0.3 to 1.4 seconds, the shortest were overestimated and the longest underestimated: this led directly to the concept of an *indifference point* or *indifference zone* corresponding to a duration for which there is no systematic error.[3]

---

[3] The duration of a stimulus is said to be overestimated when the latter is estimated in some way (mostly by reproduction) to be longer than it really is. In this case we shall speak of *absolute overestimation*.

As regards the duration of this indifference zone, the research carried out, particularly in Germany, in the second half of the nineteenth century produced very questionable results. Woodrow (1934) pointed out that this zone was fixed between 0.3 and 5 seconds by these authors. We shall show in the following paragraphs that this value can be modified by various factors, but it is still fixed at approximately 0.6–0.8 second by all those authors who have carried out really serious research on this point. Wundt (*Éléments de psychologie physiologique*, 1886, vol. II, p. 322) gave 0.72 as the value and his pupils, Kollert, Estel, and Mehner, gave durations varying from 0.71 to 0.75 second (quoted by Woodrow, 1934).

In our opinion, Woodrow's determination is the most accurate (1934). He used a great number of subjects, obtaining only one value from each. Despite this they overestimated short intervals and underestimated long ones. The indifference zone was found to be between 0.59 and 0.62 second, according to the method of calculation. Still based on the same tests, an interval of 0.3 second was found to be overestimated by up to 6.2 percent, the underestimation of an interval of 1.2 seconds, 2.1 percent, and that of an interval of 4 seconds, 4.6 percent. We have established a systematic deviation of between +19.6 percent for an interval of 0.30 second and −3.9 percent for 1.5 seconds (Fraisse, 1948). When the durations are filled, the values and systematic errors observed for the indifference zone are still more or less the same (Stott, 1935).

Apart from these systematic errors, estimations checked by a reproduction method may also be observed to vary with the duration. This variation averages 10.3 percent for all subjects for a duration of 0.20 second (standard deviation from the mean of

---

There is said to be overestimation *of the duration of a stimulus A* in relation to the duration of a stimulus B when A is estimated to be greater than B whereas the two durations are physically equal. This implies that A appears to be equal to B when B is physically longer than A. We shall call this *relative overestimation*.

Relative overestimation cannot be deduced directly from *absolute overestimation* (or vice versa), for in the second case we are comparing a perceptual response with a physical stimulus and in the first we are comparing two perceptions.

the reproductions). When the duration increases, the error decreases systematically down to 0.6 second (7.8 percent), then increases again as the durations become even longer (10.1 percent for 2 seconds) (Woodrow, 1930).

We shall endeavor to show later that there is a relationship between the indifference zone and specific physiological processes. Its value may vary, however, according to the conditions under which perception takes place. The most important phenomenon which tends to modify its value is the development of a *central tendency* related to the range of durations perceived in a given situation. In the course of everyday experience, we form absolute impressions which correspond to the central tendencies for apprehended stimuli. We say a chair is light or heavy according to our experience of the average weight of a chair. By virtue of a law of economy, we spontaneously expect a stimulus to be of around average value and we tend to minimize small differences—law of assimilation—or to overestimate them if they are fairly large—law of contrast (Fraisse, 1947). The result is that, if we judge the value of a range of stimuli, we overestimate those which are below the average and underestimate those above it. This law is particularly well illustrated by temporal intervals. Hollingworth (1909) showed that the divergence of results obtained by different authors in the determination of the indifference zone bears a relationship to the range of durations they used in their experiments. This can be confirmed by experiment: we used two ranges of stimuli on the same subjects, with a reproduction method, and found that the indifference point appeared at 1.14 seconds for stimuli of between 0.2 and 1.5 seconds and at 3.65 seconds for stimuli from 0.3 to 12 seconds (Fraisse, 1948).

At this point we should also mention the anchoring effect; that is, the influence of a reference value on the estimation of the duration of other stimuli. If subjects are first asked to judge durations of between 0.25 and 1 second according to a scale of 5 points (from very short to very long) and they are then asked to do this again, each perception being preceded by a reference duration of slightly more than a second, there will be an increasingly marked drift of their subjective scale, in that they judge durations

to be long and very long more often than before. This phenomenon is particularly noticeable in the case of the longest duration,
which is nearest the reference duration (Postman and Miller,
1945). There is therefore an effect of assimilation, which has also
been observed by Goldstone, Lhamon, and Boardman (1957)
and by Weinstein, Goldstone, and Boardman (1958). There
would have been an effect of contrast if the reference duration
had been far longer than the durations being estimated. If the
reference value is halfway between the values of the durations,
it facilitates the formation of a central tendency and acts as an
indifference zone, in relation to which the shorter durations are
overestimated and the longer durations underestimated (Philip,
1944). The order of presentation of the stimuli can also influence
our impression of the brevity or duration of the intervals. Benussi
(1907) found for a series of intervals ranging from 0.09 to 2.7
seconds, that the indifference point of his subjects' judgments
(from short to long or vice versa) changed position, appearing at
0.23 seconds when the durations were presented in increasing
order of length, at 1.17 seconds when the order was reversed, and
between 0.58 and 0.72 with random order.[4]

This could mean that the interval of 0.6–0.8, known as the indifference zone, is relative to the range of perceptible durations.
If this is the case, the fact that different authors have obtained
the same results would be explained by the fact that the zone
around 0.70 corresponds to the *central tendency* for habitually
perceived durations, which range from approximately 0.1 to 1.8
seconds for a single interval.

It cannot be doubted that the existence of a central tendency
is of some importance; this explains why the value for the indifference zone is modified according to the range of durations
used in an experiment. It seems, however, that the zone around
0.70 second corresponds to a specific physiological process, for it
is to be found in connection with phenomena of various different

[4] The perception of duration can also be influenced by the attitude of the
subject, as can that of any stimulus. Woodrow found that a subject underestimated an interval of 0.6 second by 0.198 second when he was instructed
to fix his attention on the limits, to be regarded as a pair of sounds, and by
0.283 second when instructed to direct his attention toward the duration of
the interval between the sounds.

kinds where no true central tendency is observed.

We shall endeavor to make a systematic study of this zone and thereby arrive at some more general hypotheses concerning the nature of perceived intervals.

2. The Zone Around 0.70 Second

Having found that the most accurately reproduced interval was of around 0.75 second, Wundt went on to compare this with other phenomena having the same duration. This is approximately the time necessary for the apperception of a complex number of 5 to 6 digits and also for the association of two words.

"From this we may conclude," he wrote, "that ¾ second is approximately the speed at which processes of association are most easily accomplished; consequently, in reproduction, we try involuntarily to make objective spaces of time equal to this speed, by shortening longer intervals and lengthening shorter ones." From this fact he went on to the following hypothesis: "Surprisingly enough, this interval is almost the same as that taken for the swing of our leg when we are walking quickly. It seems not unlikely that this psychic constant for the mean duration of reproduction and the most accurate estimation of intervals has developed under the influence of body movements, which are the movements most frequently exercised, and which have determined the tendency we have to organize and give rhythmic arrangement to large spaces of time." (Wundt, *ibid.*, vol. II, p. 322.)

Guyau took up this same idea: "Even today we still adapt the speed of our representations to the rhythm of walking and it is a natural tendency to adapt the pace of time to the pace of our thoughts and our steps." (*La genèse de l'idée de temps*, 2nd ed., 1902, p. 94.)

Wundt's supposition is confirmed by numerous facts observed in research of a very different kind. The duration of 0.75 second seems to be a psychic constant corresponding to the duration of the complete process of perception. The study of behavior gives simple evidence of this which concurs with psychophysiological facts. Let us look into this.

As regards behavior, the efficacy of a sensation appears to

reach a maximum when it precedes the reaction by about 0.75 second. We have already seen that the optimum interval between conditioned and unconditioned stimuli is between 0.5 and 1 second (Wolfle, 1930; Bernstein, 1934; see Chapter 2, pp. 50-51). If the conditioned stimulus consists of a combination of two factors, conditioning is most easily established when there is an interval of about 1 second between them (Czchura, 1943).

In a completely different field, research on the psychological refractory period has shown that the temporal interval between two signals must be at least 0.5 second for the response to each signal to be equally quick. If the second signal is given too soon after the first, its response is delayed; Welford (1952) considers that this delay of central origin is due to the second stimulus being held in reserve until the centers are free to use it and produce the second response. This phenomenon is observed even if the subject gives no response to the first signal. The time of reaction to the second signal is at a minimum when the first signal precedes it by about 0.6 second. This is true whether the first and second signals are of the same nature or not, provided that the duration of the interval between the two signals is varied at random (Fraisse, 1957, 1958).

To give a more straightforward example, subjects who are asked to reproduce a sound stimulus by pressing a button begin to react about 0.7 second after the stimulus has ceased, as if this interval were the optimum for immediate succession (Oléron, 1952). There have been attempts to make a more direct evaluation of the duration of the perceptual process—or of apperception, as Wundt calls it—by measuring the time necessary to give a response to a stimulus. In experiments on *reaction time*, where the motor response must be given *as soon as* the stimulus is perceived, we do not actually measure the duration of the whole process of perception, for the response is very automatic and occurs right at the threshold of perception. There is, therefore, a space between the moment of perception of the presence of a stimulus and the moment of identifying it. This is shown by experiments carried out with a tachistoscope, where an interval is easily perceived between the beginning of the system which leads to the production of an excitation (release of a shutter, drop of a screen, contact) and the moment when its contents are perceived. Wundt tried to measure this interval by the method of simple reaction time, but he failed, for it is impossible to know the phase in the perceptual process to which the response of the subject corresponds (quoted by Woodworth, *Experimental psychology*, pp. 305-306).

Experiments in reaction time involving selection are more helpful for they require identification of the stimulus; in this case, however, the duration of the selection of the response must be taken into account as well as the time of perception. Obviously the results obtained in such experiments vary with the experimental conditions. It is interesting to note, however, that this variation ranges between about 0.3 and 0.6–0.7 second, the latter duration being reached when discrimination, while still perceptual, becomes difficult. This point is illustrated by an experiment by Lemmon (1927). The subject is placed in front of two frames, one for each hand. A certain number of bulbs can be lighted on each frame and the subject must react with the hand on whichever side the more lights appear. The average reaction time for distinguishing between one light and none is 0.29 second; between 2 and 1, 0.475 second; between 3 and 2, 0.566 second; between 4 and 3, 0.656 second; between 5 and 4, 0.741 second.

Approximately the same time is required for the identification of simple stimuli, checked by giving the name or number. Cattell (1885) found that about 0.40 second was necessary for reading letters, simple words, or numbers of 2 digits. When the subject has to indicate the number of dots in a given area as soon as possible, he takes 0.42 second for one dot and 0.63 second for 5 (see Szelinski, quoted by Woodworth, *ibid.*). If he is instructed to concentrate on accuracy, he takes between 0.60 and 0.70 second to count 1 to 3 dots (Jensen, Reese, and Reese, 1950). These experiments all show that perceptual identification is a process which requires a duration of between 0.30 and 0.60 second, including the time necessary for the response. But the response is rapid in this case; that means that it corresponds to the *beginning* of perceptual identification. When speed of response is not insisted on, the time which elapses between the moment of stimulation and that of the response is often between 0.60 and 0.80 second. The time necessary for the identification of simple stimuli, according to Colegrove (1898), is 0.6 second; the same value was also found by Ross and Fletcher (1953) in a test concerning the perception of colors, when subjects with normal color vision were asked to identify the stimuli.

From these facts alone it is already possible to infer that the complete process of perception lasts about half a second. Of course there can also be a response to the *presence* of a stimulus; it is this duration (0.15 to 0.2 second) which is measured in simple reaction time. There may also be a response as soon as there is *identification* (after 0.3–0.4 second). Beyond this, a decline in the process probably takes place; if a new stimulus occurs just at the end of this process (i.e., after about 0.6–0.7 second), there is not only duality of perceptions, but the second perception seems to take place immediately after the first with no intervening gap but also without any overlap. This seems to be the most acceptable interpretation.

Attempts have been made to check this duration of the perceptual process by more direct methods.

Calabresi (*La determinazione del presente psichico*, 1930) used an ingenious method. We are capable of apprehending— without recourse to memory—a total of 7 or 8 letters. However, if this number of letters is presented to us for a very short space of time (10 milliseconds), we perceive on an average only 4.1 letters. This number is more or less independent of the total number of letters presented; if only 4 letters are presented they are almost always all perceived. Therefore if two groups of 4 letters are presented rapidly with a sufficient interval between them, we perceive about 8 letters. What would happen if two groups of 4 letters were presented in rapid succession at the same point (with Wundt's two-shutter tachistoscope), each for 10 milliseconds? Calabresi found that the number of elements retained varied with the interval, as follows:

0.05 sec.	4.4 letters
0.2 sec.	5.5 letters
0.4 sec.	5.6 letters
0.7 sec.	6.5 letters
1.0 sec.	7.2 letters
1.2 sec.	7.2 letters

An interval of about 1 second is therefore necessary before the two perceptual groups are almost integrally apprehended. If the interval is shorter, the two processes may be considered to interfere with each other; the first is not yet finished when the second begins. From this we can deduce that the complete perceptual process requires between 0.75 and 1 second.

Other authors have used more analytical methods based on simple perceptions. They try to measure the duration of the establishment of a sensation and the duration of its decline. The values found by this method obviously apply only to touch, hearing, and vision, for which senses the duration of the peripheral processes themselves is not enough to invalidate the calculations. The duration of the process includes the latent period, which varies with the intensity of the stimulus. The value of this period at its irreducible minimum is between 30 and 70 milliseconds for auditory sensations and between 70 and 110 milliseconds for luminous sensations. But these are the lower limits. Moreover, the latent period only measures the time necessary to reach the absolute threshold. From this moment the evolution of the process continues. The sensation continues to grow, even in the case of brief stimuli; the primary cortical excitation brings about associative cortical reactions which are necessary for perceptual recognition (Piéron, *La sensation guide de vie*, 3rd ed., 1955, p. 461). The duration of this phase is difficult to measure, but obviously it grows with the complexity of the stimulus and indirect reference values help us to estimate its approximate length. For instance, the reaction of cessation of the *alpha* rhythm occurs only 0.1 to 0.9 second after the beginning of sensory stimulation. Gastaut (1949) interprets this interval as the time necessary for the attention to be directed toward the image projected onto the "screen" of the visual cortex, while Hebb (*Organization of behavior*, 1949, p. 71 ff.) estimates that the duration of reverberating cortical circuits, which he considers as the basis for the process of perception, is about 0.5 second.

Following this phase of the establishment of perception there is a phase of decline and disappearance. Piéron (1935) estimates the duration of the latter at between 0.15 and 0.2 second for luminous sensations; Von Bekesy (1933) found durations of around the same length for auditory sensations, while Buytendijk and Meesters (1942) estimated that the duration of this decline is even longer.

It is difficult to determine the exact durations of the various phases of the process of perception, but the psychophysiological facts are clear: the average duration of this process must be around half a second. It would be preferable, however, if we could measure it by more direct physiological means. There are obviously great difficulties in the way of this, but the way indicated by Gastaut (1949) gives us some useful indications. If the potentials resulting from a light stimulus are recorded directly on the cortex, this stimulus is found to induce, in both humans and animals, a very complex cycle of excitation which lasts, according to Gastaut's tracings (*"L'activité electrique cérébrale,"* p. 68), between 0.5 and 0.6 second.

These facts prove that a brief stimulus brings about a process of greater duration, which is manifested on a physiological, perceptual, and motor level. By relating these data, we can try to interpret the observations from which we started out. It seems that one perception appears to succeed another without transition when it occurs just at the end of the process of perception of the preceding stimulus. If it were to begin before the end of the latter, the overlapping of the process of decline of the first and the process of establishment of the second would lead to the perception of contiguity or of an *integral pair*, which is typical of intervals of less than 0.75 second. If, on the other hand, the second process begins after the end of the first, we perceive separation and the two processes cannot be associated without an effort on the part of the subject, which probably gives rise to a supplementary process of association. This effort is still only effective within certain temporal limits, which are those of the perception of time itself.

If the processes are contiguous, the effort involved in distinguishing clearly between the two successive stimuli leads to overestimation of these intervals; if they are separated, the efforts of associating them leads to underestimation. These primary determinations are influenced by other stimuli which occur at the same time and bring about the phenomena of the central tendency and of anchoring. These reinforce or counteract the effect of the former.

These deductions explain the hypothesis proposed by Wundt. The optimum duration for association is that at which one process is finished when the next begins. The association with the duration of one step in walking, and we would add that of a heartbeat, does not mean that one of these rhythms controls the others. In any case, they have different constants according to Arrhenius' law (see Chapter 1, p. 33).

It is far more likely that all these phenomena correspond to an optimum rhythm for successive associations in the nervous system. Walking, heartbeats, movements effected at a spontaneous tempo, and perceptions all follow on at intervals of about 0.70 second, which we consider to be the optimum interval for the functioning of the nervous centers because it is the most economical.

II *Perceived Duration and Physical Changes*

When analyzing the conditions of succession, we showed that they determine the quality of durations. Our task is now to reconsider in more detail the relationship between perceived durations and the duration of the physical changes which gave rise to them. We shall consider the following points in turn: (1) The perception of empty time; (2) The perception of filled time; (3) Filled time and empty time; (4) The duration of continuous changes; and (5) Differential sensitivity.

1. The Perception of Empty Time

We have already explained the concept of *empty* time as opposed to filled time (Chapter 3, pp. 77-78). In the nineteenth century the idea of empty time was only meaningful from the point of view of the psychology of experienced content. Time was empty when no sensations were produced. Now everyone agrees that *empty* and *filled* are not terms that can be applied to perception but describe the *physical* situation: it is in this way that we shall use the words.

Theoretically there are two cases in which empty time may

exist: (a) A duration with a background of vague and ill-defined sensations is delimited by two brief stimuli (auditory, visual, or tactile); (b) The duration corresponds to the cessation of a definite stimulus (e.g., interruption of a sound or a light). The first case has been studied more than the second, which is ambiguous. Indeed this cessation, from the perceptual point of view, is either like an indistinct background, the interruption of a stimulus which in itself is characterized by duration, or it is the figure against a ground constituted by the continuous stimulus; this second case brings us back, both in theory and in practice, to that of the perception of filled time, or the perception of continuity.

In the last section we considered the perception of empty time: we saw that these durations are overestimated when they last less than 0.75 second and underestimated when they are longer, thus implying the existence of an indifference point which may vary with experimental conditions.

It now remains for us to study the influence of the different types of stimulus which can delimit empty time. In Chapter 3 it was shown that we do not perceive plain duration but the duration of an organization of stimuli. We must now show that our perception of the duration of an interval depends on the nature of its limits, which cannot be dissociated from it. As we shall see, the relationship between interval and limits is complex. Sometimes the limits are incorporated in the duration, through the action of assimilation, and sometimes, through contrast, the interval appears to be isolated from its limits. These effects may be different according to whether the time is more or less than 0.75 second. The processes of organization of the limits are more or less facilitated by the nature of the stimuli or by the attitude of the individual. For example, if we hear three sounds following one another at regular intervals and we aim to perceive two of them as a group and the other isolated, the interval between the grouped sounds will appear shorter than the other (Benussi, *Psychologie der Zeitauffassung*, 1913, pp. 115-117).

a. The Sensory Nature of the Limits. When the stimuli are of equal physical duration, the longer the corresponding sensory processes are, the longer the delimited interval appears. In general we perceive as one unit interval and limits; the longer the latter last from the physiological point of view, the longer the duration of the whole seems to be. If the limits are tactile or auditory, the duration seems shorter than if they are visual (Meumann, 1893). Here again we see the difference between sensations with rapid processes and those with slow.

b. *The Intensity of the Stimuli.* In the case of brief durations, the more intense the stimuli (auditory) the shorter the interval seems (Benussi, *op. cit.*, p. 335); the duration of the process of perception of the first sound, which is longer if the stimulus is more intense, seems to *eat into* the interval which follows it. This is particularly noticeable if a series of stimuli with identical intervals is used instead of two sounds. The more intense the stimuli the *denser* they appear to be and hence the intervals seem briefer (Meumann, 1894). If the durations are longer the phenomenon is far less marked, which is understandable since the duration of the sensory processes becomes negligible in relation to the interval.

The interval is shortened in the same way if the first stimulus is more intense than the second; the same explanation applies as above. On the other hand, if the second stimulus is more intense the interval seems longer, the duration of the final process coalescing in some way with that of the interval. This is true only for short durations (Benussi, *op. cit.*, p. 335).

This influence of the intensity of the stimuli should be compared with that of the intensity of taps when the interval is not perceived but produced. If a subject is asked to reproduce an empty interval by pressing a Morse key twice, the travel of the key being constant but the resistance variable, the interval produced is found to be shorter when the effort necessary for the signal is greater. This result means that the interval delimited by the more intense efforts is overestimated by comparison with that delimited by less intense effort (Kuroda, 1931).

c. *The Pitch of Sounds.* Intervals delimited by sounds of higher pitch appear longer than those of lower pitch (Triplett, 1931; the pitch varied in his experiments from 124 to 1024 cycles per second).

The greater the difference in pitch of the limiting sounds, the longer the interval seems to last (Benussi, *op. cit.*, Cohen, Hansel, and Sylvester, 1954). This effect may be counteracted, however, by consonance of the limiting sounds.

This question has not been studied much, despite its interest for music, but it would appear that the greater the consonance, the easier the organization of the limiting sounds and the shorter the interval will seem.

d. The Duration of Sounds. When the duration of the limiting sounds is increased, the apparent duration of the interval between them is also greater. If one of the sounds is long and the other brief, the interval between them is overestimated when the long sound comes first and underestimated when it is second. In the first case, it is incorporated to some extent in the interval; in the second, the end of the interval coincides with the beginning of the final sound (Woodrow, 1928a).

e. The Position of Empty Time. An empty interval may have a number of different positions in the context of a perception. Benussi (*op. cit.,* p. 411) studied the influence of the duration of anticipation on the perception of an empty duration and obtained the following results: if the signal precedes a brief empty duration by 0.45 second, underestimation ensues; if it precedes it by 3.15 seconds, there is overestimation. On the other hand, this duration of anticipation has no influence on long durations. In other words, a brief period of anticipation promotes the impression of very short duration and a long period does the reverse.

The results found by Israeli (1930) were different: short periods of anticipation, from 0.18 to 0.54 second, gave rise to overestimation of the subsequent empty intervals (duration from 0.35 to 1.09 seconds) and this overestimation increased the smaller the empty interval. The divergence of the results obtained by Benussi and Israeli may be attributable to their methods, but also to the attitude of their subjects, which definitely plays an important part. Schumann (1898, showed in a fairly rapid succession of 3 sounds a—b—c that, if b—c is shorter than a—b, it appears even shorter than it really is, for c arrives, as it were, too soon in relation to our expectation, based on the duration a—b. If on the other hand b—c is greater than a—b, c arrives too late and b—c is overestimated. These conclusions are true if a—b—c are perceived as a group, but if a is considered as a signal for the interval b—c, the phenomenon no longer occurs.

Still according to Israeli (1930), a sound which follows an empty interval is overestimated. The same is true, as Benussi had already shown, if the empty interval is delimited before and after by two stimuli: the effect of overestimation in this case reaches a maximum. It is a sort of transcription into temporal terms of the Müller-Lyer illusion.

2. The Perception of Filled Time

The general law of overestimation of short durations and underestimation of long ones is as valid for *filled* durations as for *empty* ones, although the former case has been less studied (Edgell, 1903; Anderson, 1936). Starting out from this law again, we can study the effect of the different forms of "filling" on the apparent duration.

a. Divided Intervals. Between the case of *filled* duration and that of *empty* duration there exists an intermediate case: where the interval between two limits is filled with discontinuous stimuli. This phenomenon is analogous to that of dots in space. We know that a divided interval appears longer than an empty interval of the same duration (Oppel's illusion), but does this apply to time as well as space?

On the whole, results confirm that it does, but we should first note, as Bourdon (1907) does, that this kind of evaluation is very difficult. To compare two intervals which are both divided but to different extents, or to compare one divided interval with one empty or filled interval, is the same as comparing two *forms* of different quality. Most authors agree, however, that: (a) a divided interval appears longer than an empty interval of the same duration; (b) this effect is lessened when the total duration of the interval increases for the same number of interpolated sounds; and (c) an interval with more divisions appears longer than one with fewer (Hall and Jastrow, 1886; Wundt, *op. cit.*, vol. II, p. 323; Münsterberg, 1889; Israeli, 1930).[5] Furthermore, of

[5] All these authors obtained their results by a method of comparison; Wirth (1937) found the same law by a method of production. His subjects were instructed to produce a divided interval equal to a reference empty interval, by tapping. The interval produced was shorter than the reference interval, showing that the produced interval (divided) is overestimated in relation to the empty reference interval.

two divided intervals, the one which is evenly divided appears longer than that which is irregularly divided (Grimm, 1934).

The difficulty of the comparison lies in the fact that it is not easy to judge the duration of the divided interval while ignoring the impression of speed given by the succession of elements. Some subjects may base their judgment on the number of sounds produced, in which case the divided interval will appear longer, while others will be content to estimate the speed of succession, in which case the divided interval may well appear shorter. This is perhaps the reason for Benussi's statement that divided intervals appear shorter than empty intervals (*op. cit.*, p. 483).

A recent experiment has given us the answer to this problem (Fraisse, 1961). The subjects were instructed to estimate, by reproduction, the time it took for a series of black dots, travelling at different speeds, to cross a window which was narrow enough that only one stimulus was seen at a time. We found that the variable speed had no influence. On the other hand, when the speed is constant, if the subject is given a reference with a certain frequency and instructed to reproduce an equivalent duration by releasing stimuli at a different rate, the estimated duration will be longer when the frequency is higher; this is true for visual stimuli (passage of black dots), and for sounds or tapping at different rates. The results were as follows (reference 10 seconds):

Stimuli		
Dots Passing Window	Reference 2/sec. Reproduction 6/sec. 7.0 sec.	Reference 6/sec. Reproduction 2/sec. 8.5 sec.
Sounds	Reference 1/sec. Reproduction 3/sec. 7.5 sec.	Reference 3/sec. Reproduction 1/sec. 10.5 sec.
Tapping	Reference 1/sec. Reproduction 3/sec. 8.1 sec.	Reference 3/sec. Reproduction 1/sec. 11.2 sec.

Speed must therefore not be confused with frequency.

b. The Sensory Nature of the Stimuli. Auditory and visual

stimuli from 1 to 16 seconds are reproduced identically whatever the ambient conditions (Hirsch, Bilger, and Deathrage, 1959). This is confirmed by the work of Hawkes, Bailey, and Warm (1960), who found, by three methods (reproduction, production, and verbal estimation), that auditory and visual stimuli and electrical stimulation of the skin, with a duration of between 0.5 and 4 seconds and comparable subjective intensity, were estimated with the same accuracy, the last mentioned perhaps appearing to last a little longer. When using the method of absolute judgments, however, Goldstone, Boardman, and Lhamon found that visual stimuli lasting 1 second were judged to be longer than the personal standard of 1 second more often than were auditory stimuli. But this is mainly due to the relative intensity of the two.

c. *The Intensity of the Stimuli.* A more intense sound will seem longer than a less intense sound. This law, which, in the physiology of sensations, governs very short excitations, is also applicable to perceptible durations. The effect decreases, however, as the duration increases (Oléron, 1952).

The same law was established by Hirsch, Bilger, and Deathrage (1956) under more complex conditions. The subjects were instructed to reproduce visual or auditory stimuli lasting 1, 2, 4, 8, and 16 seconds, the conditions of their presentation being variable. Their perception was not affected by the presentation and reproduction of the stimuli in the dark or in the light, but when stimuli were presented in a silent environment and reproduced with accompanying noise, the reductions were longer than was the case for stimuli presented with noise and reproduced in silence. The authors interpret this fact as evidence of a relationship between perceived time and the level of auditory stimulation.

This law is probably also the explanation for the fact, found by Van der Waals and Roelofs (1946), that the duration of presentation of an object seems longer when the latter is larger and more complex.

d. *The Pitch of Sounds.* A high-pitched sound appears longer than a deep one. Research has been carried out with comparisons ranging from 128 to 1024 c.p.s. (Triplett, 1932) and 1000 to 3000 c.p.s. (Cohen, Hansel, and Sylvester, 1954b).

3. Filled Time and Empty Time

Does a filled duration appear longer than an empty one? Referring mainly to Meumann (1896), most authors affirm that, when physically equal, filled durations appear longer than empty durations. But Meumann himself observed that this is true only if the empty interval follows the filled. We shall see (p. 146) that the relative position in time of the two durations to be compared is a source of systematic errors which make all the results obtained by this method very unreliable.

Using the reproduction method, Triplett (1931) found that certain subjects overestimated empty durations while others did the same for filled durations; we have observed ourselves that there is no significant difference when subjects reproduce empty and filled intervals of 0.5 and 1 second. This applies to children as well as adults (Fraisse, 1948a).

Triplett also used the two forms of empty duration: the interruption of a stimulus and an interval delimited by two stimuli. These results are also very variable. Some subjects actually perceive the interruption of the sound as a figure against a ground.

Doehring (1961) found that there was no difference in the accuracy and reliability of reproductions of temporal intervals between 0.5 and 8 seconds whether the intervals are empty (the subject presses a Morse key twice) or filled (he continues pressing for the whole reproduction).

Such unstable results make it unnecessary to consider the details of certain hypotheses. There has been much discussion concerning the possible reasons for the relative overestimation of filled time; it has usually been thought that such time attracts more attention and hence is overestimated. This hypothesis expounded by Meumann adumbrated the phenomenon of centration which Piaget so rightly stressed. However this may be, there is still the problem of *attitude* which may vary from subject to subject, with the relative position of the stimuli, the instructions, and the method of estimation (Curtis, 1916).

4. The Duration of Continuous Changes

In all the cases so far discussed, we have been dealing with the duration of a stimulus which remains identical or that of an interval delimited by two stimuli of the same nature. We should also consider the perception of the duration of a change. There are two different cases, as Piéron points out. In the first case there is a change in the *position* of the stimulus, or movement. In the second, the stimulus changes in quality or intensity: a color which changes gradually or a sound which swells.

What are the laws governing the perception of the duration of these changes?

a. The Influence of Space on Duration. If two stimuli are situated a certain distance from each other in space and succeed each other in time, it will be observed that the apparent duration of the temporal interval increases to a certain extent as the distance increases.

This fact has been confirmed for visual space (Abbe, 1936, 1937; Cohen, Hansel, and Sylvester, 1953) and for tactile space (Suto, 1952, 1955).

The task of the subjects in practical experiments is to compare two temporal intervals delimited by three successive stimuli. If the distance between the second and third stimuli is greater, the temporal intervals being equal, the second interval will seem longer. When the ratio of the distances was 1:10, Cohen, Hansel, and Sylvester found an effect of 12 percent. The effect on the forearm is apparently much the same (Suto, 1955). Cohen, Hansel, and Sylvester suggested that this should be called the *kappa* effect, to distinguish it from and relate it to the inverse interaction which has been named the *tau* effect by Helson and King (1931). Following Benussi (1917), these two authors showed that the distance between two successive tactile stimuli depends on the temporal interval separating them. The greater the latter, the longer the apparent distance.

The interaction of space and time (*kappa* effect) in the field of touch is directly connected with the visual perception of space, as has been shown by Suto (1955). He found that when we close our eyes, the comparison of two temporal intervals delimited by three points of excitation on the forearm is accompanied by visual images. This test was repeated on blind people, who had lost their sight in early childhood, and it was found that they did not have this illusion, which the Japanese call S. As blind people do not spatialize the stimulated points, their estimation of time does not vary with distance. Sudo (1941) and later Suto (1959) found that, in the case of vision, the effect of space does not depend on its physical but on its apparent extent. In Sudo's experiment, the three stimuli were situated on the apices of the terminal angles of a Müller-Lyer figure and the spatial illusion influenced the apparent duration of the two temporal intervals. Suto confirmed that when the distances are equal there is relative equality of the temporal intervals. The S or *kappa* effect is also influenced by the direction in space (Cohen, Hansel, and Sylvester, 1955). It reaches a maximum when the direction is downward, and a minimum when it is upward; for horizontal directions the effect is intermediate. According to these authors, the effect caused by the experience of speed of movement according to the direction in space may have some influence as well as distance: acceleration downward, the reverse upward, and uniform speed on the horizontal plane.

b. The Effect of the Speed of Change. What effect does the speed of a change have on its duration? An ingenious experiment carried out by Brown (1931*b*) proved that the greater the speed, the shorter the duration seems to be. He passed in front of the subject a strip of paper, on which a small figure was drawn. This ran between two lateral screens which limited its apparent travel. The subject was instructed to regulate its speed until the duration of the figure's appearance seemed equal to the duration of the temporal interval between two auditory or visual signals. Brown then increased the apparent speed of the figure simply by reducing the illumination. When the speed is equal, the apparent speed is in fact greater if the appropriate area is less brightly illuminated (Brown, 1931*a*). He then found that modification of the apparent speed caused a modification of the apparent duration such that the equation time $= \dfrac{\text{distance}}{\text{speed}}$ was experienced phenomenologically. Cohen and his colleagues (1955) compared

Brown's results with their own *kappa* effect. They observed that the duration appears longer if the space between two stimuli is increased; Brown obtained the same result by diminishing the speed. In both cases the results point to the general law of an inverse influence of distance and speed on time; Cohen saw in this the effect of everyday experience.

Obviously, however, the perceptual estimation of duration in the experiments we have just quoted are not constructions or deductions. As Koffka (*Principles of Gestalt psychology*, 1935, p. 296) so rightly said, the experience of time depends on every factor in the field of perception. Between space and time there is coalescence and reciprocal interaction (*kappa* and *tau* effects). The size of one influences that of the other, in much the same way as the terminal angles of the Müller-Lyer illusion influence the apparent length of the line they delimit.

The part played by experience in the effect of distance and speed on the perception of duration can only be clearly shown in developmental research. The experiments carried out by Piaget (*Le développement de la notion du temps chez l'enfant*, 1946) and ourselves (Fraisse and Vautrey, 1952) have shown that when young children compare the duration of the motion of two objects moving at the same time, they sometimes find that the object which moved more quickly was moving for the longer time. (See Chapter 8.) We can deduce nothing from this to help our present problem, for the duration of the movement is somewhat too long to be perceived and it also refers to two events taking place at the same time. We can only infer from this that it is not certain that the relationship between perceived time and speed or distance is the same for children as for adults.

The same relationship between duration and speed can also be seen in changes in quality or intensity. The apparent duration of a sound of increasing intensity becomes briefer as the speed of increase grows (Fraisse and Oléron, 1950). When two intervals are delimited by sounds of rising pitch it also seems that the smaller the difference in pitch, the shorter the apparent interval (auditory *kappa*). (Cohen, Hansel, and Sylvester, 1954.)

The results found by C. O. Weber (1926) may be compared with these cases. If a subject has to execute a movement of given duration (this being indicated by two successive sounds) and a given length, the duration produced increases with the effort involved (this experiment is carried out with Michotte's kinesimeter with different weights added). This means that the interval produced is underestimated as the effort produced increases.

5. Differential Sensitivity

This question was the object of much study in Germany during the nineteenth century, following the impetus given to psychophysiology by Fechner. The results were, however, very disappointing owing to the absence of any standardization of methods and the small number of subjects used. The work carried out during the past 30 years has, however, given us more accurate and reliable results.

Two main methods have been used: comparison and reproduction. In the first case the standard method of calculating the threshold is to find the difference in duration between the constant stimulus and the variable stimulus; this has as much chance of being perceived as not. In the second case a dispersion index of the reproductions is used.

Goodfellow (1931) showed that the results obtained by these two methods were very close, provided that the necessary precautions were taken. This is true only *on an average* however, for the correlations of the results obtained by different methods average only .50 in the case of auditory stimuli. Individual differences in attitude probably have something to do with this.

We shall now consider in turn the values of the threshold in the cases which have been most thoroughly studied.

a. Empty Time (Auditory). Using the method of comparison, Blakely (1933) found that between 0.2 and 1.5 seconds the threshold is lower than 10 percent, with a minimum of 8 percent for durations of 0.6–0.8 second. It increases from 10 to 16 percent when the duration is increased from 2 to 4 seconds, and to 20 and 30 percent for durations from 6 to 30 seconds. Mach (1865) found a minimum of 5 percent for 0.4 second, Goodfellow 6.5 percent for 1 second. These results concur; they also coincide with those found by the reproduction method. Woodrow (1930) determined the threshold (standard deviation from the mean) for a wide range of durations and obtained the following values:

0.2 sec.	10.3%
0.6 sec.	7.8% (minimum)
1.0 sec.	8.6%
2.0 sec.	10.1%
4.0 sec.	16.4%
from 5 to 30 sec.	from 16 to 17%

Woodrow had the same interval reproduced 50 times in succession. We have obtained higher threshold values: 12 to 14 percent for durations from 0.2 to 1.5 seconds and 12 to 20 percent for durations from 0.3 to 12 seconds, when the subject does not know in advance the duration of the interval to be reproduced (Fraisse, 1948c).

Inversely, the threshold can be lowered by systematic training: Hawickhorst (1934) found a threshold of 3.6 percent for an interval of 1 second after training. Renshaw (1932) even found an average variability of 1.2 percent in the reproduction of the duration of a second after training 5 subjects for 159 days.

b. *Filled Time (Auditory)*. The differential thresholds are very similar to those of empty time according to Blakely (1933) and Stott (1933, quoted by Woodrow, 1951).

c. *Empty and Filled Time (Visual and Tactile)*. In the case of the duration of a light stimulus, Blakely (1933) found very similar results to those obtained for auditory stimuli. These results concur with those of Hulser (1924), according to which, for the evaluation of the duration of a stationary point of light, the threshold is 10.3 percent for 0.75 second, 6.5 percent for 1.55 seconds, and 5.4 percent for 2 seconds, and with those of Quasebarth (1924) who found a threshold of 7 percent for a duration of 2 seconds for the presentation of a stationary spot of light, and 14 percent for 8 seconds.

Goodfellow (1934) made a close study and comparison of the differential thresholds, for the duration of 1 second, of intervals delimited by auditory, visual, and tactile stimuli. He combined the results obtained by three methods of measurement (constant stimulus method, method of just noticeable differences, and reproduction method) and found the following thresholds:

For hearing	from 7.0%
For touch	from 9.5%
For vision	from 11.5%

Thus the differential sensitivity is approximately the same, although a little less acute in the case of vision; this is easily explained by the nature of the limiting stimuli. The differences may also, however, be partly due to the fact that we are not accustomed to judging the duration of visual or tactile stimuli. This interpretation is supported by Gridley (1932). He used the test with which Seashore measures the time sense, by comparing two successive intervals delimited by two pairs of sounds, but then went on to replace the sounds by tactile stimuli and compare the results. He found slightly lower values for touch (72.8 percent success) than for hearing (77.8 percent success). But if the same test is repeated, the results for sound improve by 2 percent and for touch by 4.1 percent. It is therefore possible that the differences observed may be much reduced by practice.

d. The Weber-Fechner Law. Nineteenth-century study of the perception of time was dominated by the problem of whether Weber's law was applicable to time. Fechner was emphatically in favor of this point of view but results differed so much from one author to another that Nichols came to the conclusion in 1890 that Weber's law could not be applied to temporal intervals. This opinion has been supported by Bonaventura (1929) and more recently by Maack (1948) and Woodrow (1951).

There are several preliminary points which must be considered. Essentially Weber's law applies to the relative differences in *intensity* of the various stimuli. Fechner of course tried to generalize the law, while Kiesow (1925, quoted by Piéron, *op. cit.*) thought that it might be merely one particular case of a more general law according to which we are only sensitive to relative differences. Even in this wide sense, the law is confirmed only for variations in the intensity of one and the same stimulus and not for variations in quality.

As regards the perception of duration, we have seen that the distinction between "more and less" results in considerable qualitative differences between the intervals. We can therefore expect Weber's law not to apply perfectly in this case.

No distinction is made in any of the works we have referred to between perceptible durations of up to 2 or 3 seconds and longer durations which can no longer be perceived.

What can we conclude from these facts? If we refer again to the figures given in sections *a* and *b*, especially those of Woodrow, we find that the relative differential threshold remains more or less constant between 0.2 and 2 seconds: it varies only from 7 to 10 percent. For all these results there is one value for which we have optimum sensitivity; this generally coincides with the duration of the indifference zone, i.e., around 0.75 second. But there is nothing exceptional about this, for in every case where Weber's law applies the differential fraction reaches a minimum and then increases when the intensity becomes lower or higher.

Certain German authors were under the impression that the relative differential threshold varies in accordance with a periodic law; that is, that there are several minimum values for sensitivity (Estel, Mehner, and Glass).[6] But the methods they used and their small number of subjects and measurements had already been criticized by Fechner. This phenomenon has not been observed again in any recent work and it is now only of historical interest.

The relative constancy of the differential fraction was probably not observed owing to a failure to distinguish between perceived durations and durations which were only estimated. This is indicated by the following fact: as we shall see, Weber's law may be applied fairly well to estimated durations (Chapter 7, p. 213), but in this case the differential fraction is far larger, around 20 percent. This is illustrated by Woodrow's results (1930); between 2 and 4 seconds the value of the differential fraction changes, corresponding, in our opinion, to the change which

[6] In this connection see Nichols (1890) and Bonaventura (*Il problema psicologico del tempo*, 1929) who favor this point of view to some extent and think there are 3 minimum values for sensitivity: 0.35–0.40 second; 0.70–0.80 second; 2.15–2.5 seconds.

occurs between the process of apprehension and that of estimation.

If we consider this question from the point of view of Fechner's law, we may also ask whether perceived durations form a continuum which bears a relationship to the logarithm of the physical durations. This problem is difficult to tackle. Edgell (1903) used Plateau's method of equal distances which consists in finding a value for a stimulus such that the differences between it and two other stimuli, one larger and one smaller, appear to be equal. When Fechner's law is confirmed, the value for this stimulus approaches the geometrical mean of the other two stimuli and not their arithmetical mean. But Edgell found that the estimation of a duration halfway between two others corresponded more or less exactly to their arithmetical mean. It may be objected that the task to be performed by the subjects is very difficult and their results are therefore variable and questionable.

A few years ago Stevens (1957) suggested the substitution of an exponential law for Fechner's logarithmic relationship; he tried to establish this by seeking the sensation which appeared to be half or double another. By means of successive approximations, other authors have even established subjective scales of intensity; that is, levels of sensation which correspond to different sizes of stimuli. Units have been proposed for the measurement of sensations by evaluating one in relation to another; thus we have the *sone* for the perceived intensity of sounds (Stevens), the *veg* for weight, the *gust* for taste, the *dol* for pain, the *bril* for perceived luminosity.

Gregg (1951) tried to establish a subjective scale of durations along the same lines. He aimed to find the (filled) durations which appeared to be half of stimuli of 0.4, 0.8, 1.6, 2.4, and 4.8 seconds. When he offset the errors due to temporal position, he found that the average estimations of his subjects corresponded (more or less) to a stimulus of half the duration (some subjects systematically overestimating and others underestimating). This result corroborates that which Edgell obtained by another method. In view of this, it is difficult to understand why Gregg still proposed a subjective time scale in which the unit would be the *temp,* one *temp* corresponding to a second. A *half-temp* would correspond to a duration of 0.50 second (to within 5/1000), two *temps* to a duration of 2 seconds.

Ross and Katchmar (1951) tried the same thing for empty durations. They also found that, on the whole, the estimated half of a duration corresponds to the physical half, allowing for errors due to position which they disregarded. They proposed another unit, the *chron*, corresponding to our experience of a duration of 10 seconds. Ekman and Frankenhaeuser (1957) claimed that they had found an exponential law (with a power of 1.55) for durations from 1 to 20 seconds. But their results are questionable. The temporal position error plays a large part in the establishment of these temporal scales where it is not possible to compare the reference with the reproduction with relative simultaneity, whether the latter is equal to or half the former.

In their results the reproductions of equal length are appreciably shorter than the reference but this is less apparent in the case of the reproduction of half the duration; this points to an exponential relationship. There is also the anchoring effect to be considered; the shorter durations are overestimated and the longer underestimated.

The most recent work on this subject (Björkman and Holmkvist, 1960) is very satisfactory from the point of view of method and the results concur with those of Gregg. The subjects had to set, by turning a knob, a sound equal to or half of a reference (durations from 1 to 7 seconds were used) and they could hear the reference and their setting as often as they liked; this greatly reduced the temporal position error. Under these conditions, for an interval between two sounds of 0.1 second, they found the following values for the equal settings S_1 and for the half settings $S_{\frac{1}{2}}$:

Reference	S_1	$S_{\frac{1}{2}}$
1.0 sec.	0.921 sec.	0.410 sec.
2.5 sec.	2.047 sec.	1.028 sec.
4.0 sec.	3.253 sec.	1.612 sec.
5.5 sec.	4.472 sec.	2.311 sec.
7.0 sec.	5.784 sec.	2.919 sec.

Apart from a slight underestimation in the settings, the value of the apparent half is truly half of the apparently equal setting. These authors obtained the same results as Ekman and Frankenhaeuser using the reproduction method, but the same criticism applies. It seems, therefore, that there is little to be gained from subjective time scales since, allowing for difficulties connected with the method, the apparent half of another apparent duration is equal to the true half of the latter.

6. The Influence of Attitude

All the laws which we have so far discussed apply to the central tendencies of groups of individuals. Very little research has been done, however, in which the authors did not find marked individual differences which have even gone so far as to reverse the phenomenon (e.g., underestimation instead of overestimation). Many results even seem to contradict one another in this respect.

It is essential to recognize the importance of the attitude of the subjects in the perception of time. This has become clear in every field of perception, starting with the work of the Würzburg school which was recently confirmed on a more experimental basis by Bruner and his colleagues.[7] Our perceptions are a function of the nature of the stimuli, but also of the "assumptions" with which we apprehend them. This assumption itself depends on our previous experience, on the context of the perception, and on our personality, all these factors contributing toward our attitude. The less compelling the event, the more difference our attitude can make. Not only does it have an effect on our constant selection of sensory information and on the significance we attach to this, but it can even modify the apparent size of objects. This has been proved by Bruner's and Goodman's experiment on the overestimation of the size of coins by poor children.

It is permissible to think that attitude plays a larger part in time than in space, for all perception of succession is evanescent by its very nature, whereas in the case of spatial perceptions, it

[7] For an account of this research, see Fraisse (1953).

is possible for the perception to be compared with its object.

No doubt, therefore, there is some justification for explaining divergence of results by differences in the attitude of the subjects, but this factor has not been checked systematically enough, even by using different instructions. We do not even know the exact variation possible in attitude as regards the estimation of duration. It seems right to make allowance for the degree of organization of successive sensations. This organization may depend on the nature of the stimuli, but it also may depend on the attitude of the subject. If, in an experiment involving reproduction, a subject is asked to pay attention to two limiting sounds and to reproduce them as a pair of sounds, the interval he reproduces will be much shorter than if he is told to listen to the limits passively and direct his attention toward the sensations which fill the interval (Woodrow, 1933). To give another example, following Benussi, it has been suggested that the more attention we pay to an interval, the more it is overestimated. This fact became apparent in the comparisons between two successive intervals. Benussi thought that this was the explanation for temporal position errors. Of two successive, equal intervals, the one on which we fix our attention appears longer. Quasebarth (1924) confirmed this interpretation by telling his subjects to listen passively or actively to one of two successive intervals which were to be compared.

This law should be compared with Piaget's work on perceptual centration. In the course of his study of the genetic development of perceptions, he was led to assume that the stimulus on which vision was centered was overestimated in comparison with peripheral stimuli. He subsequently generalized this law of centration and extended it to any stimulus which predominates in a perceptual comparison (because it is the reference or yardstick). We have confirmed that these effects of centration correspond to a centration of attention: that is to say, they correspond to an orientation toward a stimulus, the centration of our vision being only one particular instance of a more general phenomenon (Fraisse, Ehrlich, and Vurpillot, 1956).

This centration may originate in the object, which draws the attention of the subject, or in the subject, whose attitude creates an orientation. In the case of time, centration may be induced:

1. By the nature of the task. For instance, comparison usually causes centration on the second interval which is more present in our minds when we make our judgment; hence the fact that the temporal position error is generally negative owing to over-estimation of the second element;

2. By the difficulty in perceiving time, e.g., when the limits pertain to different sense modalities;

3. By some feature of the duration which particularly attracts our attention, e.g., the greater intensity of one limit (empty time) or of the continuous stimulus (filled time);

4. By the instructions, which suggest that a temporal interval has more or less absolute or relative importance.

Is there a relationship between the overestimation of intervals on which we center our attention and the fact that we judge time to be longer when our attention is fixed on the duration more than on what is enduring (we shall discuss this law in Chapter 7)? Benussi tried to establish this. Although we cannot definitely prove it, we think that he was wrong, for the conditions of perception of brief durations are generally very different from those of the estimation of long durations. Moreover, these over-estimations of the duration on which the subject's attention is centered have occurred most often in experiments where they had to compare two durations; that is, when they were mainly concerned with the *duration* of each interval. Why should they pay more attention on the one hand to the changes themselves, and, on the other, to the actual duration?

To sum up, the perception of duration is a function of our attitude, of which the most important element seems to be the attention paid to perceived time. The greater this attention, the longer the interval seems.

But, all things being equal as regards attitude, the perception of duration depends on the nature of the changes perceived. These bring about different processes of excitation at the receptor level and, at the central level, perceptual processes whose organization determines the apparent duration.

PART III

||||||||||||||||||||||||||||||||||||||||||||||||||||||||||

CONTROL OVER TIME

||||||||||||||||||||||||||||||||||||||||||||||||||||||||||

Man has a great advantage over animals. He is able to form representations of changes other than those he perceives in the present.

Through these representations he can embrace, from his present, the past and future time perspectives which form his temporal horizon (Chapter 6).

The interval between the present moment and a future satisfaction makes us aware of duration through emotional reactions. These feelings of time help us evaluate the duration, but our estimations are more generally based on the quantity of changes we notice in it (Chapter 7).

The representation of changes leads to representations of successions and durations; when these are interrelated they give rise to a notion of time which becomes more and more abstract as we grow older. Man is then capable of relating all sequences of change and all temporal intervals independently of his immediate experience. He can thus gain control over change within the limits permitted by its irreversibility (Chapter 8).

The distinctions we suggest between the different modalities of our control over time are justified by developmental psychology; children can have a temporal horizon and feelings of time and can appreciate duration before they conceive a notion of time.

But the full richness of our temporal horizon and the greatest possible accuracy of our evaluations of duration are not attained until, through the notion of time, we become capable of reconstituting all change.

|||||||||||||||||||||||||||||||||||||||||||||||||||||||||||

THE TEMPORAL HORIZON

WE LIVE IN THE PRESENT; IN OTHER WORDS, OUR BEHAVIOR IS A function of everything which determines it *here and now*. But these present activations are constantly referring us to what has already passed away or to what has not yet come to be.

The present, therefore, has several dimensions: ". . . the present of things past, the present of things present and the present of things future" (Saint Augustine, *Confessions*).

In this ever-changing world, our actions at any given moment do not only depend on the situation in which we find ourselves at that instant, but also on everything we have already experienced and on all our future expectations. Every one of our actions takes these into account, sometimes explicitly, always implicitly.

To put it in another way, we might say that each of our actions takes place in a temporal perspective; it depends on our *temporal horizon* at the precise moment of its occurrence.

Before commencing our study of this question, it would be useful to define its extent by showing the difference between the temporal horizon of man and the vague horizons of the animal world on the one hand, and, on the other, the notion of time arrived at by an adult in full possession of his intelligence.

In one sense animals do have a temporal horizon. They seem to live only in a world of perceptions, and nothing is more present than a perception. But every perception is a signal and this fact in itself refers them to the past. A stimulus only acquires significance through previous experience, when a connection has been established between the stimulus—which has become conditioned—and the reaction.

The perception-signal also orients activity; is it not true that this always seems to have a purpose? The conditioned stimulus determines anticipation behavior: seeking food, fleeing danger, etc. The rat, which hoards food, acts as if it were able to foresee famine; this behavior is instinctive, but it is based on the past, for those animals which have been most deprived of food are those which hoard most (Morgan, Stellar, Johnson, 1943).

Animals do not, however, refer explicitly to past events in their behavior; nor do they have a purpose. Their temporal horizon is purely implicit. They act "as if . . ." Man also sometimes behaves like this, but he is also able to make conscious use of all the dimensions of time. On the one hand he can recall the past as such, that is to say he *can retell a story;* he recognizes it as having belonged to his experience in the past. This recognition is never complete unless he can locate the memory. On the other hand he can organize his activity with reference to *projects* which are representations of the future.

Stories and projects differ from pure imagination in that they always contain a temporal reference to previous or subsequent changes which locate them in relation to the present experience.

We can thus represent the past or the future without having a "representation of time." Here we should draw a fine but necessary distinction between the temporal horizon and the notion of time. The idea of a homogeneous background which underlies our notion of time (Chapter 8) is a different thing from the representation of one or more past or future events. The representation of an event takes on a temporal aspect from the moment it is placed in relation to others. The most simple case is obviously that of reference to the present, the special moment which determines the two sides of our experience. More often we locate events in time by placing them in relation to each other. We thus arrive at the formation of temporal perspectives, just as there are perspectives in space. Thus when we represent the past or the future, we do not have some abstract scheme of time in mind but a series of events which are ordered in accordance with planes of succession.

The representation of succession is not complete until we have acquired a notion of time; as we shall see later, this is the only way in which it can take into account the different natural series of events and their intervals. But this notion of time is not necessary for the formation of the temporal horizon; this is proved by the fact that the latter already exists in children at a stage of their development where they are not yet capable of operational constructions.

In its first stages the temporal horizon is simply a manifestation of memory, and it develops with the latter. This chapter is not, however, devoted to a specific study of memory but to a study of the perspectives of the temporal horizon itself; it deals mainly with the way in which we can behave in relation to the three aspects of time: the past, the present, and the future.

I *The Nature of The Temporal Horizon*

The temporal horizon develops slowly through childhood. Its beginnings therefore present us with an excellent opportunity to start this analysis by establishing its *nature*. We shall study the quantitative development of temporal perspective with age in the second part of this chapter.

1. The Beginnings of Temporal Perspective

At birth a child is only capable of more or less diffuse reflex reactions which give its behavior a distracted nature (Malrieu, p. 26).[1]

A reflex reaction is one which follows immediately on a stimulus. The first cry of a baby is a reflex reaction to the entrance of air into his lungs and sucking movements are occasioned by anything touching his lips. There is no temporal perspective in such reactions. But through the classical action

[1] In view of the thorough work of Piaget (*La construction du réel chez l'enfant*, 1937) and Malrieu (*Les origines de la conscience du temps*, 1953), we shall attempt no more than a broad outline of the initial developments of the temporal horizon.

of conditioning, the first temporal references soon appear: the "practical series" of which Piaget speaks (*op. cit.*, p. 325). From , the very first weeks of his life, a baby who is hungry will stop crying when he is lifted up to be fed and his mouth will seek his mother's breast before it touches it (Piaget, *ibid.*, p. 326; Malrieu, *op. cit.*, p. 37). We have already shown that babies adapt very rapidly to the rhythm of feedings and even to the omission of one feeding during the night.

This conditioning implies temporal seriation; there is anticipation of the future at the same time as practical utilization of past experience. One stimulus becomes the signal for another: at two months, for instance, a baby will turn his head in the direction from which he has heard a noise (Piaget, *op. cit.*, p. 326). This utilization of signals implies the existence of a temporal horizon, but at this age it develops entirely on the plane of what is being lived; the past and the future are simultaneously present in his behavior of the moment. Thus reaction chains are gradually formed in which each event becomes the signal for the next one. The way in which a child helps his mother dress him from the age of about ten months shows that he is becoming capable of adaptation to more complex temporal series. Later he will even take the initiative by performing the first gestures of a series, guided by more long-range anticipation; thus he will go and find his coat and shoes so that he can be dressed to go out.[2]

In all these initial instances of temporally organized behavior it is the past which gives meaning to the stimulus, turning it into a signal, but the signal gives rise to behavior which is oriented toward the future. At first it is the very near future (the baby seeking his mother's breast), but it gradually becomes more and more distant (the child who goes to fetch his coat so that he can go out for a walk). Thus at first the future manifests itself as an orientation "towards," like an attitude of seeking, but we can see how this orientation will gradually come to be accompanied by an actual representation of the anticipated satisfaction or the danger to be avoided.

[2] Malrieu (*op. cit.*, p. 58) regards this behavior as make-believe, but we think it is of the same nature as that of the baby who stretches his arms out toward his mother to be picked up.

In the second stage, which appears after the first but does not replace it since the two develop simultaneously, operant conditioning now becomes apparent, which involves temporal perspectives. In the case of operant conditioning, the child must discover, as animals do, the action which will give him satisfaction. To quote another example given by Piaget (*op. cit.*, p. 334), after seeing a rattle the baby must learn to pull a cord in order to move it. When he does this, he reinforces an association which was found by chance through groping and which was fixed through the effect it produced. In the case of such reactions, the goal cannot be reached until it is "the wished-for future which organizes the present." (Malrieu, *op. cit.*, p. 60.) The child must reconstitute a useful succession, disregarding the goal itself for a moment in order to make the gesture necessary in accordance with a before and an after.

In cases of simple conditioning the succession is experienced, but here it is reconstituted. As the child develops, these successions become increasingly complex. It is no longer simply the goal which organizes the reaction, but the memory of it. The child's actions are oriented toward an object which is no longer present, at least in his field of vision. The reaction is delayed. Let us take the example of an 18-month-old child who is in one room and suddenly decides to go and look for a toy in another room. This behavior implies a memory which is located in time and space, but this memory only acts as a promise of future satisfaction. Thus the past and the future are found, at least initially, to be relative to each other. Piaget (*op. cit.*, p. 336) found a first example of this delayed reaction in the following behavior of a baby of 8 months:

"Laurent sees his mother come into the room and follows her with his eyes until she sits down behind him. He then goes back to his toys, but he turns round from time to time to look at her. Yet there is no sound or noise to remind him of her presence."

This obviously shows the beginnings of memory and of location in time and space.

Whether delayed or not, these reactions constitute what Piaget calls *subjective series;* they are gradually transformed into *objective series,* which differ from the former in that the apprehended succession is objectified. When faced with a situation, the child no longer simply remembers his action, but also the object itself. The corresponding behavior is to seek a toy which has just disappeared behind a screen. Every parent knows that a baby will cease to want something if it is removed from his sight, but they also know that this possibility cannot last. Gradually the object which has disappeared remains present through memory and the baby no longer stops demanding it when it is not tangibly present.

Through the effect of simple and operant conditioning, a child who is about 1 year old will have acquired a temporal horizon which is gradually becoming independent of his own reactions. "Time finally overflows the duration inherent in the actual activity; it is applied to things themselves and provides a continuous and systematic link between events in the outside world. In other words, time ceases to be simply the scheme necessary to any action, binding the subject to the object, and becomes the general background embracing both the subject and the object." (Piaget, *ibid.*, pp. 346-347.)

At this stage the child's behavior still depends on signals provided by his surroundings. Gradually, however, the perceptions which have led to actions will make the absent objects live again as memories. These may be simple representations, but they will become really individualized as the child learns to speak and gives them names; speech will also make it possible for him to act on them, combining them into series which are independent of the present action. Let us again take an example from Piaget (*ibid.*, p. 352): "Jacqueline (aged 1 year, 7 months) picks a blade of grass and puts it in a bucket as if it were one of the grasshoppers brought to her a few days ago by her little cousin. She says: *'Totelle* (= sauterelle), *totelle, hop-là* (= sauter) . . . *garçon* (= son cousin)'."[3] A perception has recalled to her a series of past events which speech permits her to name and reenact.

The effect of these representations and the horizon it opens

[3] (Translator's note.) Approximate translation: "grasshopper . . . jump . . . boy."

are also shown by the words of a child of 2 years 1 month, quoted by Decroly and Degand (1913): *"Lait, parti, Maïette,"* which in this context means: *"J'ai bu mon lait, je vais chez Mariette,"* emphasizing the double perspective, past and future, of the present.[4]

As Piaget often repeated, the first developments of speech do not strictly speaking give rise to new forms of behavior, but to transpositions on to the plane of speech of what the child was already capable of doing. He uses words as he used gestures. "Again" or "wait," probably the first words which have a temporal reference for the child, are the equivalent of a little pair of arms held out toward the mother. The child who asks for "spoon," to try to eat his food, is only doing the same thing as the baby who grasped the cord in order to agitate the rattle.

Speech will permit a considerable extension of the temporal perspectives. Through language the individual not only has all his own past at his disposition, but he can also know that of the community in which he lives. We shall see later how this horizon develops progressively with age and how the two sides of it, the future and the past, change in relative importance.

The temporal perspectives of a child develop, therefore, through his experience of series of events and of actions, prolonged by the memories of the individual or the group. It would be wrong, however, to think that a child escapes from the domination of more or less reflex reactions, elicited by present stimuli, and from the distraction they lead to, simply by training or through intellectual constructions. The temporal horizon of a child develops at the same time as the unity of his personality. The latter requires that the child learn to inhibit the reactions triggered by his body or surroundings, particularly his emotional reactions, in order to be capable of taking into account what has gone before or what is to follow. Only the attainment of emotional stability can permit him to undertake actions which have more far-reaching effects or refer to a more distant past, for emotivity tends to enclose us in the present. In this respect

---

4 (Translator's note.) "Milk, gone, Maïette" . . . "I have drunk my milk and now I am going to see Mariette."

we can say with the psychoanalysts that time develops for a child in so far as the pleasure principle gives way to the reality principle (Bergler and Roheim, 1946).

2. The Constitution of the Past

In his early experiences a child learns to recognize temporal series in which one element permits him to foresee the following one. But these series are not yet located in relation to the present of the child. They are also ambiguous by their very nature. They are at the same time the manifestation of a past experience and an orientation toward a future experience: an animal which seeks food or a child who reaches for his bottle or his mother's arms is referring at the same time to a past and a future simply by recreating a temporal series. As Heidegger notes, the future is in a way an accomplishment of the past; it presupposes the past, but only has the sense of past if there is a future (Biemel, *Le concept du monde chez Heidegger*, 1950, p. 124).

Temporal series therefore do not in themselves create temporal perspectives in which there is a distinction between the past and the future. They only enrich present experience. This fact is well illustrated by the language of a child. Whether he uses verbs or not in his sentences, he speaks only in the present until the age of three. The context also shows that he is essentially expressing a present situation even if there is a reference to the past. If he says "Mommy gone" he means that Mommy is not there and not that she went out a few hours ago. The same phenomenon is to be found in the evolution of languages. "All primitive languages express the idea of *action* by verbs but they do not all make a clear distinction between the various tenses. In its primitive form, one verb form can be used to designate the past, the present or the future." (Guyau, *La genèse de l'idée de temps*, 2nd ed., 1902, p. 6.) In actual fact this primitive language expresses the reality of the objective world and not a temporal experience. "The future and the past have a sort of pre-existence and eternal survival in things themselves; the water which will flow by tomorrow *is* now at its source, the water which has just flowed by *is* now a little further down the valley. What is past or future for me is present in the world." (Merleau-Ponty, *Phé-*

noménologie de la perception, 1945, p. 471.) Time "is born out of my relationship to things" (*ibid.,* p. 471); in other words, time is born as soon as things bear a relationship not only to each other but to the subject of the experience. For this to be, the events we experience must first become memories. I can only relate to myself that which has an existence, and memory is the means of keeping present what has passed away.

Not everything in our past experiences is transformed into memories, however. A large proportion is not fixed. There is a considerable discrepancy between the immediate richness of a perception and what we can recall of it only a few seconds later. The deficiency is not uniform and there is actually no correlation between the richness of the content of a perception and the amount which is transformed into memories (Fraisse and Florès, 1956). In the first place "we only retain what has been dramatized by language." (Bachelard, *La dialectique de la durée,* 1936, p. 58.) We have to be able to name things, people, and feelings for them to join our memories. This is a necessary condition, but it is inadequate by itself; they must also be integrated in some way with other memories. Without the creation of this relationship, it is not possible to recall them.

Whatever the conditions of the transformation of experience into memory, its fixation alone is not enough to give it a place in our temporal horizon. Memory is not "the integral and passive recording which many authors have thought it to be, as if it were only necessary to consult the register of one's memories and one would find the pages in the correct order together with a table of contents corresponding in advance to every possible classification." (Piaget, *Le développement de la notion de temps chez l'enfant,* 1946, p. 260.)

This is easy to see in a child. When he is three or four years old he is content to place all his memories in a single moment which he calls *yesterday.*

"When a child of between 2 and 4 years old wants to retell the story of a walk, a visit to some friends or his adventures on a journey, a multitude of 'juxtaposed' details tumble forth incoherently; each one is associated with others in couples or little successions, but their overall order escapes the habits of our mind." (Piaget, *ibid.*, p. 261.) Children also fail in arranging in order a series of pictures which form a story; this shows well that it is not only words which fail them. Even adults find it difficult to reproduce the order of memories which do not constitute a natural or logical series. If, for instance, we read four unpublished poems to some students and, without warning them, ask them the next day to recall the order in which they heard them, four out of five subjects are incapable of replying correctly. Our memories do not automatically bear a relationship to one another, as if there were some sort of sedimentation or a phonographic recording, as Guyau puts it.

All authors are agreed today that memory is a construction. In this construction the recency of a memory is only one factor, and at that it is the weakest. We do not recall memories by beginning with the most recent, which should be the most intense, and moving back toward the oldest and dimmest. We locate our memories in relation to one another by trying to remember the order in which we actually lived them. What cues do we use for this reconstruction?

To begin with, every event has a sort of "temporal sign." In our memory every action is associated with all the circumstances which accompanied it. Certain of these may help us to date the event in some way. They bear a relationship to the basic elements of the calendar; that is, with the succession of morning and evening, the rituals of eating and sleeping, the succession of the days of the week, several of which have their own particular features (Sunday, payday, etc.), the succession of holidays, months, seasons. These concomitant circumstances give us specifically temporal signs which take on their full significance when they are part of a conceptual framework, but originally they have the nature of experience; we may even think that they are closely linked not only with changes in the outside world but also with the rhythm of our own organic changes (this is confirmed in our study of temporal disorientation, p. 163). We saw in Chapter 1 that organic changes take place in synchronism with the different phases of the day; they are connected with the succession of meals, and of activity and sleep, and anticipate these. These successive states no doubt also confer some kind of temporal sign on all that we do, although we are not conscious of them. This is the opinion of Kleist (1934): he believes that temporal location is dependent on the midbrain centers. This idea is taken up by Delay (*Les dissolutions de la mémoire*, 1942, p. 136): "It is likely that this region constitutes the real clock of the organism because on it depend all the main periodic rhythms (hunger, thirst, sleep, sex needs). These periodic vegetative occurrences have a temporal constituent which may act as a basis for the general chronological registration of the impressions we experience."

These temporal signs give individuality to memories; but they are not sufficient to organize them in such a way that they constitute temporal series. This is where true construction comes into play, using every means available to give events an order in relation to each other. Paradoxically enough, space is very helpful. Our actions usually take place in successive surroundings. Thus space, by imposing an order on our actions, becomes a means of reconstituting their true succession in our memory. My memories of the Spanish towns I visited this summer did not fall into a spontaneous order in relation to each other, but I can easily find my route on a map and thus locate them in time.

This recourse to space is only one particular instance of a more general law: we use our knowledge to place our memories, in fact we seek the most probable order of events. Groethuysen (1935-36) noted that when retelling the events of the day soon afterwards, we follow the chronological order, but several days later we present the facts in the order in which they should have taken place. If we recall a meal, we do not put the dessert before the hors-d'oeuvre. The most easily reconstituted associations are those which correspond to a causal relationship. Piaget stressed the influence of the apprehension of causal relationships on the development of temporal series. The term "causal relationship" must be understood in its widest sense, including successions determined by what Ribot called the logic of feelings. Research concerning testimony shows that the recounting of experienced events is a reconstruction in which the deepest tendencies and interests play a part.

Finally, our temporal horizon is constituted by the organization of our memories. This organization may be founded on the temporal cycles which give individuality to memories, but it makes systematic use of the guiding thread of necessity to reconstitute the temporal order of past events. Thanks to the effect of these organizations, our temporal horizon can develop far beyond the dimensions of our own life. We treat the events provided by the history of our social group as we treat our own history. In fact the two tend to merge; for instance, the history of our childhood is that of our first memories but also that of the memories of our parents, and this part of our temporal perspective develops from both sets of memories. These are both better organized thanks to the clocks and calendars provided by society, which give us the cues that are indispensable when it comes to long periods of time. Klineberg (1954) quotes the case of Indians in California who did not know their own age nor how long ago an event took place which dated from more than 6 years ago.

The importance of logical construction in the constitution of our temporal horizon is shown by the fact that the temporal horizon of mental defectives, like that of young children, is very limited. Both are incapable of assembling their memories to form a past (and of anticipating the future); they are prisoners of the present. De Greeff (1927) estimated that in extreme cases of mental deficiency the temporal horizon of the past does not exceed about 10 days. This is the maximum duration ascribed by them to a past event (the end of the war, the time during which they have been at a special school). Beyond this point everything is on one plane, for they are incapable of ordering their memories.

In dreams, when there is no logical arrangement of memories with a view to their relation to reality, the temporal horizon is also very disturbed; we recall events which really happened, all mixed up with the fantasies of our imagination. There is no respect for chronology; we may dream of the funeral of a friend and the next moment be chatting with him over dinner. There are frequent "flash backs." The cinema often makes use of this technique, but in that case the spectator fills in the gaps or reconstitutes the logical order by not permitting himself to be drawn into the order of events he experiences. In dreams, on the other hand, we are subjected to apparently chaotic images whose principle of organization does not relate to the order in which they were experienced and its requirements, but to the effect of associations where affective preoccupations predominate. This is how we interpret Freud's theory of the intemporality of the unconscious: "The processes of the Unconscious system are *intemporal*, that is they are not ordered in time, they are not modified by the passing of time, in fact they bear no relation to time. Relation to time is linked with the workings of the conscious system." (Freud, *Das Unbewusste*, 1915, quoted by Bonaparte, 1939, p. 73.) "The sense of reality and the sense of time are both apparent in the system of Perception-Consciousness alone. The unconscious knows nothing of them: the intemporal unconscious in which the secondary process dominated by the reality principle is not yet master, the unconscious which has remained entirely governed by the primary process guided solely by the pleasure principle." (Bonaparte, *ibid.*, p. 100.)

The organization of our memories is necessitated by their confrontation with reality; that is, with the whole of our other memories, our knowledge, and the information from our present situation. In dreams or delusion, man is absorbed by the present image, which he does not connect with any other; Guyau noted the same fact, in the language of his time, by saying that it was the perception of *differences* which created time. "One thing is remarkable, the perpetual metamorphosis of images which, when it is *continuous* and without definite contrasts, eliminates the sense of duration. . . . Because of this absence of contrast or differences, considerable changes can take place without being apprehended by our consciousness or being organized in time." (*Op. cit.*, pp. 18-19.)

The above analysis becomes clearer if we consider the pathological disintegration of memory. This is seen most distinctly in the case of Korsakov's syndrome. The chief characteristic of this syndrome, from our point of view, is amnesia; this is not always generalized and amnesia of most recent events predominates. "The patient may talk about past events without making mistakes, but not remember what has just been said to him or what he has just done; he asks for some object when he has it in his hand, he wants to eat when he has just left the table, to be put to bed when he is already there, etc." (Régis, *Précis de psychiatrie*, 6th ed., 1923, p. 350.) This inability to fix present memories obviously leads to temporal disorientation in the past, since nothing or hardly anything is retained except from before the illness. Perception is normal in people suffering from this disorder, and sometimes it is transformed into a memory, but the few memories which do become fixed remain as "isolated fragments" (Jaspers, *Allgemeine Psychopathologie*, 1920). The patient may recall buying a pair of shoes, but be incapable of locating the purchase in time. It is like a book, he says, which I have inside me but cannot find (Cohen and Rochlin, 1938). The disorder seems to be manifested by a difficulty in establishing new relationships between occurrences. Experiments have shown that these patients have a selective difficulty in associating pairs of words (Ranschburg, 1939) and more generally in forming new associations (Wechsler, 1917). In particular they are unable to associate any memory with all that happened before, during, and after the event in question. This has been called *fixation amnesia* because, on the plane of experience, a new fact no longer recalls a former event, although the patient is still capable of associating ideas or of reasoning. In our opinion, Van der Horst (1956) was more correct in speaking of a loss of the temporalization of memories due to the fact that each of them remains isolated. According to Jaspers (1920), there is a "reduction in the integration of actions." A patient of Bonhoeffer's was very much aware of this; she said that "the memory of the succession of events in time is completely lacking." (Quoted by Van der Horst, 1956.)

These disorders obviously result in the temporal disorientation of the subject. They are not of an intellectual nature, for the patients make good use of objective cues, clocks, or calendars. They lack the ability to date their experiences by reference to others. This reduces their temporal horizon. One patient was, for instance, capable of recalling the principal events which had taken place during his 29 years in hospital, but he estimated that this period of his life had only lasted 2 or 3 years. He gave his age spontaneously as 32 or 33, although he was capable of recognizing the fact that he was 59 according to the calendar (Bouman and Grünbaum, 1929). We must deduce that his memories left so many gaps and were so dissociated from one another that they caused a reduction of his temporal perspective.

By analyzing Korsakov's syndrome we are thus able to determine more precisely the nature of the organization through which our temporal horizon develops. A psychic continuity is created, for our memories more or less date themselves. This *chronognosia*, as Bouman and Grünbaum call it (1929), is distinct from *chronology*, or the ability to date events according to the abstract cues of the calendar. A person suffering from Korsakov's syndrome has lost the first possibility but not the second;[5] this shows that the organization through which temporal perspective is formed is not of a purely intellectual nature.

This problem is explained by the location of the trouble. Thanks to the work of Gamper (1928), which has been verified by several other authors, it has been established that this syndrome corresponds to an affliction of the mammillary bodies and neighboring centers; that is to say, of the basal nuclei. This subcortical location of the cause of Korsakov's syndrome shows that it is not an intellectual type of disorder but must be explained by the action of mechanisms closely linked with our vegetative life. There is a relationship between the mammillary bodies and the central vegetative nuclei, which play an important part in setting the main vital mechanisms in motion: hunger, thirst, sleep, sex needs. It has therefore been established that "physiologically, the reproductivity of the events of our life depends, right from the phase of its formation in the cerebral cortex, on the reinforcement of the cortical traces by a subcortical mechanism which leads from the hypothalamus to the cerebral cortex and acts on

[5] On the other hand Minkowski quotes the case of a patient suffering from the early stages of general paresis, who was capable of recounting everything he had done during the war in the order in which he lived the events, but was incapable of saying when the war began or when the armistice was signed (*Le temps vécu*, p. 13).

the cortex in a way which cannot at the moment be defined as to quality but which is very probably sensitizing." (Ranschburg, 1939, p. 531.)[6]

As regards the actual mechanism of the influence of the mammillary bodies on the integration of experiences, we are reduced to conjecture. Two hypotheses have been suggested by Delay and Brion (1954). According to the first, the bodies emit a basic rhythm which is indispensable for the development of psychological activities elaborated in the cortex; according to the second, they act selectively on a mechanism of memory by registering a temporal sign.

Thus physiopathological analysis clarifies our psychological descriptions. The periodic changes of vegetative life imprint themselves deeply on everything we experience. We become aware of this when we spontaneously relate events to the periodic recurrences in our surroundings or our activities; but this impression probably also exists on an even deeper, biological level.

The relationship between disturbances in temporal orientation and somæsthesia has been pointed out several times. Revault d'Allones (1905) quoted the case of a patient who no longer felt any need to eat, drink, or urinate and, correlatively, complained of no longer "feeling time" and not knowing what time of day it was except by using artificial memory aids as we do to date the year or the month; that is, by reasoning. Similarly Cohen and Rochlin (1938) found, in the case of the patient we have already mentioned (p. 164), disturbances in his experience of his body-self.[7]

---

[6] These observations should be compared with those recently made on patients who had had a dorsomedial thalamotomy following emotional disorders and unbearable pain. These patients, who had had neither mental deficiency nor temporal disorders, were prone to more or less transitory temporal confusion after their operations. They made mistakes concerning the time, the day, the date, the season, the time which had elapsed since they came into hospital, and the date of their operation. Like patients with Korsakov's syndrome, they showed a dissociation between the direct estimation of time and judgments supported by reasoning (Spiegel, 1955).

[7] It should be noted that temporal disorientation in cases of Korsakov psychosis can leave other methods of temporal adaptation intact. The *perception* of the duration of brief successions is not affected (Gregor, 1907), nor is the *estimation* of durations of a few minutes (Ehrenwald, 1931*e*), nor, as we have seen, is reasoning concerning time. It is essential

If, therefore, a process exists which dates events in some way by connecting them closely with the rhythms of our organic life, our past is the fruit of these complex organizations discussed above which prolong, or sometimes even supplement, the more automatic registering of our memories. Cortical lesions may affect them, especially those involving the two frontal lobes: they cause a disturbance in the "act of memoration or act of mental synthesis." (Delay, *Les dissolutions de la mémoire*, 1942, p. 133.) This is manifested by amnesia where the actual temporal loss is less evident than in the case of subcortical lesions. Any neurosis may lead to this kind of disturbance, in fact even pure fatigue can do this through a transitory or lasting inability to make the intellectual effort which permits the ordering of memories. Vinchon (1920) noted this particularly in the case of schizophrenics.

To sum up, our temporal horizon is a perspective constructed from indications provided by the temporal signs of our experience and by our efforts to organize our memories by calling on all the available principles of temporal organization.

This construction is not uniform, however. If I look at my past, my memories do not come with regularity. In this past perspective there are knots formed by crucial events—a death, success in some competition, a war—which break the continuity and play the same part as planes in spatial perspective; we locate occurrences according to whether they came before or after these breaks in our existence. The distance between these planes is also not regular. Certain periods appear far longer than others although we *know* that, according to the calendar, they were of the same duration. It was first observed some time ago that this relative duration depends on the number of memories: in retrospect a period seems longer the richer it is in memories. In the same way, the distance between two planes in a landscape seems greater when there are more landmarks.

For this reason, periods in the more recent past seem longer than other periods of objectively the same length which belong

to remember this fact, for this selective disorder confirms the specific nature of the development of our temporal horizon.

to the more distant past. Yesterday has in retrospect a far longer duration than any day in past years. In addition to this effect of perspective, there is sometimes a discrepancy between the impression of duration at the time of the experience and the apparent duration of the same period in our memory. As we shall see in the next chapter, duration at the time of the experience appears longer the more changes take place. But the number of changes we *notice* is not necessarily proportional to the number of memories we retain. The days seem very long to a prisoner because he "counts the hours," but afterwards the duration of his imprisonment may seem very short because he has few memories of these days. On the other hand a day spent going round a town or a new region as a tourist seems very full that same evening; as it leaves us with a large number of memories, we shall have the same impression several years later when we recall it.

The heterogeneous character of our temporal perspectives is therefore due to the very nature of our experiences, but it is only seen through the intermediary of the quantity of our memories. It is also due to this fact that the effects of perspective are the same for historical periods, which we have not experienced ourselves, as for our own past. The duration of the centuries of French history is relatively long because we know a lot about them. This has no relationship with the distance in time: the history of the three great centuries of Athens has a longer duration in the temporal perspective of a Hellenist than the ten centuries of our Middle Ages.

It is not, then, surprising to find that the same effects of perspective are to be observed in collective as well as merely personal representations of time. Hubert and Mauss analyzed this very clearly in regard to religious life. Time is not a homogeneous background nor a pure quantity. "The parts which appear to us to be equal in size are not necessarily equal or even equivalent: *those parts are homogeneous and equivalent which are considered to be similar by virtue of their place in the calendar.*" (*Mélanges d'histoire des religions,* p. 197.) Critical dates interrupt the continuity of time, ". . . the time in which magic and religious phenomena take place is discontinuous; it passes spasmodically. . . . Thus it is not the only, or even the main purpose of the institution of calendars to measure the passing of time quantitatively. It is not based on the idea of purely quantitative time but on the idea of purely qualitative time which is composed of discontinuous, heterogeneous parts, and which revolves incessantly on itself. . . ." (*Ibid.*, pp. 199, 299.)

The Catholic liturgical calendar gives a good illustration of this description. The four weeks of Advent condense the thousands of years of expectation of a Messiah since the fall of Adam and Eve, the six weeks from Christmas to February 2nd cover the childhood of Christ, and Holy Week has the same duration as the events it recalls.

Our representations of civil life show the same characteristics. Each year which ends has had its own rhythm marked by the seasons, the chief civil or religious holidays, the holiday season, etc., and this repetition of rhythms of collective activity does more to make the years alike than the equality of their duration, for which we trust the astronomers without being able to confirm it by our own experience.

This relationship between the representations of individual time and those of historical time has a deep significance. Halbwachs, who stressed the importance of social environment in the constitution of individual memory, observed in a posthumous article (1947) that the different chronological series of our memory correspond to the different groups to which we belong: "The time of professional life, family, religious, civil or military life, is different and has different origins."

Research on ecological time has illustrated this importance of the social group for the constitution of the temporal horizon. Bernot and Blancard (*Nouville, un village français,* 1953) made a study of a French village in Normandy where two different populations coexisted, one of peasants who had been rooted for generations in the same soil, the other of glass blowers recruited in other provinces of France; they showed that the two groups had different temporal perspectives. The peasant lives in a duration which is that of his family, and his recollections go back beyond his own personal memories. "This land was bought by his grandfather, this building was constructed by his father;" to him every path, every inhabitant recalls the past. The glass blower for his part is an immigrant, cut off from his ancestors and their work. Perhaps they were workers like him and never had roots in one place or one home. Once he has been transplanted into a new region, his own memories of his youth and childhood do not attach themselves to the new background. He is almost without a past.

Every one of us who belongs to several groups has several perspectives to his past which do not coincide. We shall see (Chapter 8) that we can pass from one set of perspectives to another by reasoning and by placing all the events in an abstract time which does not correspond to real experience. But spontaneously we cannot but see the heterogeneous nature of these different perspectives: "The relationship between time in the office, time at home, time in the street, time when visiting, is often only fixed between very wide limits." (Halbwachs, 1947, p. 6.)

Without going so far as to agree with this author that, for each of us, internal duration is divided into several streams which have their source in the groups themselves, and that "individual consciousness is only the passing point of these streams or the meeting point of collective times," we believe that it is true that our temporal perspectives bear a relationship to each of the groups to which we belong and in which both our experiences and their frames of reference have their origin. The discontinuity and heterogeneousness which we find both in the perspectives of our own history and in those of historical time are of the same nature, for it is true that we are never alone and our most individual memories are closely linked with the groups in which we live.

3. Anticipation of the Future

In the first impulses of a young child there are the beginnings of anticipation. This is to some extent indeterminate. Through circular reactions and operant conditioning, a relationship is established between the impulse and the action which brings satisfaction. This connection creates the practical series of which Piaget speaks. Thus the law of effect lays the foundation for a future, since the child—like the animal—is then capable of organizing his behavior in accordance with a succession. By executing an action or observing a signal, he anticipates—on the basis of past experience—what he will do in a moment.

The temporal conditions of this association have been established through research on delayed reactions in animals. If too long an interval elapses between a signal which has significance according to the needs of the animal and their satisfaction, no connection is established. Rats cannot learn to push a lever in order to obtain food unless the period between their action and the moment of appearance of the food is less than 30 seconds (Roberts, 1930). Monkeys can find food without hesitation when this is hidden under one of two inverted cups in front of them, provided that the interval between the moment when the food is hidden and that when they are allowed to look for it does not exceed 90 seconds (Jacobsen, 1936). These intervals represent the duration for which orientation toward the future can be maintained. Obviously they vary with experimental conditions[8] and the figures we have quoted cannot be taken as absolute values.

This kind of research is also interesting from another point of view: it shows us which nervous centers are involved in this type of behavior. It has been observed that when the frontal lobes of monkeys are removed, they become incapable of delayed responses. After an interval of 5 seconds the animal does not appear to remember anything. Its immediate memory seems to be affected (Jacobsen, 1936); but Malmo (1942) showed that in actual fact it has an increased sensitivity to retroactive inhibitions. If the animal is left in the dark between the moment when it is shown the position of the correct response (in Malmo's experiment a light indicates the lever which must be pushed) and the moment of the response, the percentage of accurate delayed responses is almost as great after removal of the frontal lobes as before.

This fact explains the behavior of anticipation which requires

---

[8] Rats can be taught to wait 3 seconds before snatching a pellet of food only if the shock which punishes the violation of this "operational tabu" follows immediately on the violation. If the shock is deferred for a few seconds, either the animal takes the food as soon as it appears despite the shock which follows, or all activity is inhibited; in either case the animal does not learn to respect the interval imposed (Mowrer, *Learning theory and personality dynamics*, 1950, chap. XV).

that the excitations which orient us toward the future should
not be obliterated by new stimulations. This orientation at the
cortical level depends only on the frontal lobes being intact, for
extensive ablations of other areas have not produced any comparable inability (Jacobsen, 1936). The frontal lobes do not
seem, however, to be the only regions involved in this behavior.
Animals which have only undergone lobotomy are also less
capable of delayed reactions and it must be admitted that a
subcortical nucleus with cortical projections may also play a part
in this behavior. According to the most recent hypotheses, the
caudate nucleus is involved (Peters, Rosvold, Mirsky, 1956).
These results obtained in experiments on animals should no doubt
be compared with clinical observations on patients who have
undergone lobotomy and who have often been found to live
more in the present than before their illness and to appear indifferent to the future (Petrie, *Personality and the frontal lobes*,
1952; Le Beau, *Psycho-chirurgie et fonctions mentales*, 1954).

In man, delayed reactions are of a different nature: through
a present stimulation he becomes capable of forming a representation of the goal to be attained. By expressing this, he fixes it
and is able to return to it more easily and after a longer period.
The prospective temporal horizon of a young child does not extend beyond a few seconds, until the future is not only experienced in the present action but can also be imagined. A true
temporal distance may then take shape between the present
wish and the goal to be reached.

At first, however, these representations are nothing more than
reproductions of series which have been experienced, and the
future is only imagined as a repetition of the past.[9] Our temporal
perspectives are not fully developed until, through symbolical experiences, we become capable of conceiving a future which is a
creation in relation to our own history. This creation itself is only
possible for those who are carried beyond the present situation by
the dynamism of their activity. Generally speaking, the future
only unfolds in so far as we imagine a future which seems to us
to be *realizable*.

[9] ". . . My past no longer projected in front of me that shadow of itself
which we call our future." (Proust.)

This is clearly shown by our attitude to death. Obviously we all know that this is the end that awaits us, but it causes anxiety in us or a form of religious behavior which, on the level of a non-transcendental, "closed" religion (Bergson), is a defense against the unknown; but, as Merleau-Ponty remarks (1947), it never becomes part of our temporal perspective, whatever our age, except for religious people who consider it as the gateway to another life. When seen as an absolute end, death is not an objective to be reached.

This example shows quite well how the future is bound up with activity. As Guyau says (*op. cit.*, p. 33): "We must desire, we must want, we must stretch out our hands and walk to create the future. *The future* is not *what is coming to us* but *what we are going to.*" It is true that there are two ways of seeing the future. In one way it is the prospect of a conquest toward which we are advancing; in the other it is the anticipation of something indeterminate, accompanied by "a feeling of insecurity and anxiety" or even anguish, this last being "the experience of life reduced to experience of the future, which makes life itself into an absolute." (Lavelle, *Du temps et de l'éternité*, 1945, pp. 276-278.) In this second case there is a passive awaiting of the future, which thus seems to come toward us.

However it is seen, the future is principally the experience of a temporal interval: the distance between "the cup and the lip" as Guyau puts it (*op. cit.*, p. 34). But this is no more the object of an immediate representation in the future than in the past. Even less so. We construct our past with memories which are determined, but our perspectives in the future always remain indeterminate, especially as some of them are not simply repetitions of the past. "The experience of the future is a purer experience of time than experience of the past, where the interval is already filled and we are less sensitive, as it were, to its container than to its contents." (Lavelle, *ibid.*, p. 279.) During the whole of our life, our future perspectives remain fairly similar to those of a child for whom the entire future is located in the indeterminate domain of *tomorrow*. We can, of course, date our projects, thanks to our schemes of time and to logical constructions, but on the plane of experience there is practically nothing but the projection of desire or fear, and from this point of view our perspectives depend to a great extent on the present state of our emotions. We feel a bit tired, and immediately our projects seem unattainable, the future seems blocked. On the other hand the state of our emotions depends on the temporal distance between the present moment and a future situation. Research on the goal gradient, carried out by Hull (1934) and Lewin, has shown that the characteristics of a reaction (speed, strength) depend on the spatial and temporal proximity of the goal. There is an approach as well as an avoidance gradient. The nearer we are to the goal, the greater the force of the reaction (Miller, 1944; see also Chapter 2, p. 59). In the case of human behavior it is easy to observe these gradients during waiting (Cohen, 1953). The effects are felt by engaged couples as the day of their marriage approaches and by pregnant women nearing the time of delivery. Lewin (*A dynamic theory of personality*, 1935, p. 88) notes that criminals condemned to several years of imprisonment have been known to escape when their sentence was nearly at an end.

The two perspectives—reconstruction of the past and anticipation of the future—do not develop under at all the same conditions. As we have seen, the past is formed through the temporal sign connected with every event we experience and through the organization of our memories into series, facilitated by the calendar and by social cues in general. On the other hand, our future perspectives depend on the possibility of escaping from a present determined by the situation or from the domination of the past. There is no future without at the same time a desire for something else and awareness of the possibility of realizing it. These two conditions involve biopsychological and sociological factors which are usually closely linked. Desire grows from an unsatisfied need, but it does not develop unless we become aware, through intermediate satisfactions, of the fact that we can fulfil this need through our activity. Otherwise the desire is extinguished through lack of reinforcement. The children who have the most success at school think more of the future and have wider temporal perspectives than those who are at the bottom of their class (Teahan, 1958). Of course success in attaining gratification depends on physical and mental health (it is said that chronically sick people learn not to have any desires); it also depends, however, on the social status of the individual and the possibilities given him through education, his profession, and his fortune.

In the investigation already described, Bernot and Blancard showed that the two populations of Nouville, peasants and glass blowers, did not have the same attitude toward either the future or the past. The peasant, rooted in his soil, keeps in mind the problem of finding a position for his children and establishing them in their turn; his future is in itself determined by the main events of his life, the harvest, the duration of his lease, etc. The glass blower has no real calendar; his life is punctuated only by his work, marked by the alternations of day and night shifts; he thinks ahead of sending his children to school, then of the first job he can find them, probably at the glass works, and that is all.

If children of between 8 and 10 years old are asked to make up stories, those invented by middle-class children cover a larger period of time than those of working-class children (Leshan, 1952). The explanation for this fact is that, in the working classes, the cycles of tension and satisfaction are much shorter and people spare themselves the frustrations which result from projects with more distant perspectives. Members of the middle class, on the other hand, can organize their lives in longer cycles and act in accordance with their projects. This does not mean that the lower classes are generally less able to bear frustration. It has been clearly shown by Ellis and his colleagues (1955) that there is no relationship in a given class between the habitual tolerance of frustration[10] and the duration of the stories invented.

Nevertheless, when the general level of tolerance of frustration is equal, it is true that the conditions in which we live incite us to avoid the frustrations arising most frequently from our *posi-*

[10] On an individual level this tolerance depends principally on emotional stability, and this stability increases with age for all human beings. Biological maturation and personality development both play their part. Thus as a child grows older he learns to bear waiting better, that is to bear a delay in the realization of a conceived action (Fraisse and Orsini, 1955). He also learns to give more and more preference to the more valuable of two satisfactions, even if he must wait to obtain it whereas the less valuable would belong to him straight away (Irwin and colleagues, 1943, 1946).

tion. This defense mechanism prevents us from desiring what we cannot attain; the "sour grapes" of the fable are a symbol of this behavior. Of several possible activities, we choose the one in which we can succeed (Rosenzweig, 1933). It is therefore normal that among people who put all their energy into immediate needs, the temporal perspectives are limited to what can be attained without delay. Thus a group of American children, from middle-class backgrounds, stated that if they were to win a prize of $2000 they would save more than half of it, whereas working-class children said the contrary (Schneider and Lysgaard, 1953). Doob (1960) found similar results for three African tribes. He asked a group of very highly educated people (university level) and a group of more or less uneducated people (0 to 4 years of schooling) the following question: "Would you prefer to receive £5 sterling immediately or £50 in a year's time?" The percentages choosing the first alternative in the three tribes were as follows:

In the highly educated groups: 14, 31, 11%
In the uneducated groups: 32, 52, 24%

This result is confirmed by the replies of the same individuals to a more general question: "Is it true to say that making plans for the future is a waste of time?" The following percentages agreed that it is:

In the highly educated groups: 17, 16, 34%
In the uneducated groups: 42, 22, 66%

Even the rhythm of the payment of wages or salaries plays its part; the workman who is paid by the day does not have the same temporally organized behavior as the middle-class employee who is paid by the month or the investor who receives his dividends or income once a year.

The future perspectives of an individual depend, then, on his capacity for anticipating what is to come. This anticipation is a form of construction determined by the individual. It borrows from his past experience but it is prompted by his present desires and fits into the framework of what he considers to belong to the realms of possibility. Some readers may contrast this analysis with "castles in the air," but in this case the imagination of the individual is playing with dreams, which he does not really locate in temporal perspective, or contact with reality is lost and the sick individual believes he is living his delusions. In neither case are these dreams really projects for the future. We shall return to this point when we discuss the influence of personality on the temporal horizon.

II *The Diversity of Temporal Horizons*

The above analyses all show that the temporal horizon of each individual is the result of a true creation. We construct our past as well as our future. It is evident that adaptation is a characteristic of this activity. Man must somehow free himself from the state of change which carries him through life, by keeping the past available through memory and conquering the future in advance through anticipation. This control over time is essentially an individual achievement conditioned by everything which determines personality: age, environment, temperament, experience. Each individual has his own perspectives. Any comparison with space at this juncture could only be misleading. Space is a collection of objects and it is these which largely determine the structure of perceived space. Time is the succession of changes, but each of these—apart from the present change—only exists for us as a memory or an anticipation; in other words they are *representations*. Space is primarily *presentation;* it is imposed on us. Time is a conquest strongly marked by the personality of the individual.

1. The Influence of Age on the Temporal Horizon

There is a different temporal horizon for every age. The genetic study at the beginning of this chapter helped us to define the actual nature of temporal perspectives by watching them take shape. We shall not return to this aspect now, but we shall follow their development both in quality and quantity through the life of the individual.

The main sign which indicates that the individual takes the past and the future into account is without doubt his activity.

Any action relates to a past and a future, but this is often not explicit and does not involve location in time. An indication of the development of a temporal horizon may be found, however, in the delayed reactions which we have already discussed. These show a living link between before and afterward. Either children or animals may be used as subjects. A desired object is hidden before their eyes in a multiple choice apparatus and they are prevented from going to find it until a certain interval has elapsed. The interval may be increased according to age without impairing the success of the results. Obviously these intervals always depend to some extent on the situation but the important fact is that they increase with the age of the subject, whatever type of problem he is presented with. Hunter (1913) used a multiple choice apparatus with three alternatives and found that the tolerated interval increased from 50 seconds at 2 years, 6 months to 35 minutes at 6 years. In connection with another problem, Skalet (1930-1931) found that an interval of a few hours at the age of 2 reached 34 days at 5 years, 6 months. As the child grows he becomes capable of taking into account in his activity the things which have preceded and which will follow: "The increase of the life space in regard to the psychological time dimension continues until adulthood. Plans extend farther into the future, and activities of increasingly longer duration are organized as one unit." (Lewin, 1952.)

It is through speech, however, that we are best able to realize the extent of retrospection and projection in children and adults. The development of vocabulary and grammatical forms gives an indication for children, together with those provided by more precise observations.

When a child reaches the age of about 1 year, 6 months, he begins to imagine objects which are not present, although whether they are in the past or the future cannot be said (Lewis, 1937). At about 2 he is capable of recalling memories from about a month ago; at 3 he already has memories from as much as a year before and at 5 these go back 2 years. When he is between 7 and 8 his past begins to overflow his personal experience; he becomes interested in what preceded it, the history of his parents and history itself (Malrieu, *Les origines de la conscience du temps*, 1953, pp. 85-87).

At first these recollections are obviously not located in the past, but between the ages of 2 and 3 the child begins to use the past participle and the imperfect tense, indicating the beginnings of temporal orientation. Awareness of the future is manifested by the anticipation of sequences of behavior. A child of about 1 year, 6 months of age, arriving home at bathtime is capable of answering "bath" when asked "where are we going?" (Lewis, 1937). But only when he is between 2½ and 3 years old does he start to make references to a more distant future; a child speaks of "this afternoon" and "tomorrow," but these expressions mean nothing more than a reference to a near but indeterminate future (L. W. and C. Stern, 1907; Decroly and Degand, 1913; Gesell and Ilg, 1943). When he is approaching 3 years of age, a child begins to speak more definitely of what he wants to do the next day, and at 3½ he begins to locate events which recur every week, particularly special days such as Sunday. At 4 appear references to the following seasons, "next summer or winter," and the child begins to look forward to important festivals such as Christmas and his birthday; these are fairly precise by the age of 5 (Gesell and Ilg, 1943; Decroly and Degand, 1913). At 8 he is capable of an interest in a story which he has not experienced and he evolves his first plans for a future outside the limits of his customary activities. He speaks of his life as an adult: "I shall get married . . . I shall be a stationmaster . . . ;" these plans are more definite at the age of 9 (Gesell and Ilg).

By studying the child's comprehension of terms which give time a precise location in relation to the present moment or in relation to the calendar, we can discover how he conquers time and how he locates himself and his past and future experiences. We have drawn up a chronology of the understanding and use of terms designating a precise location in time; for this we have drawn on the work of Stern (1907), Decroly and Degand (1913), Oackden and Sturt (1922), Bradley (1917), Ames (1946), Gesell and Ilg (1943), and Malrieu (1952). Obviously the ages are approximate; various authors sometimes disagree by more than a year. The actual order of the fixation of words is more important, as this indicates the expansion of the temporal horizon.

Recognize a special day of the week, such as Sunday	4 years old
Tell whether it is morning or afternoon	5
Use the words "yesterday" and "tomorrow" with their true meaning	5
Indicate the day of the week	6
Indicate the month	7
Indicate the season	7–8
Indicate the year	8
Indicate the day of the month	8–9
Estimate the duration:	
a. of a conversation	
b. "since the holidays"	12
c. "until the holidays"	
Give the time to within 20 minutes	12

From this table we can see the simultaneous progress of location in the past and the future, as Malrieu noted (*op. cit.*, p. 84). It can also be seen that the child becomes oriented first in the cyclic activities which bear a direct relationship to the rhythm of his existence. We know that he adapts to the cycle of every day long before he knows the special days of the week. Subsequently he conquers time through the organization into sequences of the periods of time he has experienced, he orients himself and locates one moment by its relationship to others (Farrell, 1953).

As regards conventional time, the table shows that the child's orientation develops gradually between the ages of 6 and 9; only later does he become capable of estimating durations in units of time, but this is another problem to which we shall return.

Temporal perspectives develop gradually along with mental activities. For instance, a child is only able to locate an occurrence in relation to conventional time when he is capable of relating two series of events: that which he lives and that which is imposed by society. It is not until the age of about 6, when these operations begin, that he becomes capable of this orientation. Before this he can only locate his own actions by relating them to each other and classing them simply as "before" and "after." It is therefore not surprising that authors have found a marked correlation between the comprehension of vocabulary connected with time, temporal orientation, and the results of general intelligence tests, in the case of children, adults, and mental defectives (Friedman, 1944; Buck, 1946; Brower and Brower, 1947).

Considerations of this kind tell us nothing concerning the respective roles of the past and the future in the life of a child. On this point we can still do nothing more than generalize. From the moment a child frees himself from the confusion of the past and the future it is evident that the future plays a far larger part than the past in his conscious perspectives, even though his life is obviously nothing but a repetition of what he has already done and learned. "When I go to school . . . When I am seven . . . ten . . . sixteen . . . ;" these are all variations of the same theme of conversation: "When I'm grown up." If he does turn to his past it is only for brief moments, to locate himself in relation to others, but he attaches no importance to it. In the adult on the other hand, age brings with it a progressive decline in the importance attached to what is yet to come and an increase in the importance of what has already taken place. The theme of "when I am grown up, when I am married, when I have a car" gives way gradually to the theme of "in my day, when I was a child, when I was young. . . ." Old people shut themselves up more and more in a present which they live only by reference to the past (Visher, 1947).

Therefore, disregarding marked individual differences to which we shall return later, it seems that when man locates himself in time he attaches the greater importance to the longer portion of his life, taking into account the average expectation of life, that is the unlived portion when he is young and what he has already experienced when he is old. This probably explains why in every human life there is a critical period between the ages of 40 and 50, that of the middle of life when the change from youth to old age is taking place with all the readjustments of temporal perspective that this entails.

2. The Influence of Personality

We always live in the present, but there are two ways of doing so. One consists in being coexistent with the present situation, the other in detaching oneself from it and taking refuge through imagination in the past or the future. In this case the past or the future becomes a present experience. Daydreaming, a novel, or a film are examples of this kind of situation where we live primarily in a time other than that of our present activity. At the cinema I am living in the time of the film rather than merely watching a show. We could easily find extreme examples of such transpositions in mental pathology. But without going so far, our own temporal perspectives vary continuously as regards the portion of the experience relating to the present, the past, or the future. This is at the root of what Malrieu calls "temporal attitudes" which can be seen not only through the phenomenology of experienced time but also through the study of behavior, for our actions ultimately depend on them: "According to whether we are concerned with our own personal every-day life or future, or with death, the past or the future of mankind, the scope of our actions will vary greatly." (Malrieu, *op. cit.*, p. 22.)

These attitudes become apparent in each of us as we form our temporal perspectives and our personal time. It is simple to find by analysis a great number of factors influencing these attitudes, involving our approach to reality, which depends on our temperament, and the way in which our personal history has shaped us. This personal history is itself part of a culture in which the demands of time can be evaluated differently according to the era or the civilization. Malrieu showed the opposed attitudes of the eternist philosophers, who dedicate humanity to a constant effort to fashion an unalterable model for man, and the progressist philosophers, who aim at the conquest of time which is constantly bringing new things to man. Poulet, in his penetrating *Études sur le temps humain* (1950), showed through philosophy and even more through literature how time is experienced differently from one epoch to another and how much temporal perspectives vary from one individual to another. We cannot repeat his developments or summarize them, but we shall be indebted to him for many examples in our description of the main types of attitude.

In order to define these types we should do well to determine what is the most common attitude in our civilization and what importance is ascribed to the present, the future, and the past, at least in the case of adults. Israeli (1932) carried out some investigation among students and found that for them the present was 1.2 times more important than the future and 12.7 times more important than the past. This result seems to prove that we usually live in a present oriented toward the future and have little care for the past. Some more recent research by Farber (1953) gave similar results. He asked American students to number the days of the week in order of preference (1 being the favorite day), and obtained the following results:

Mon.	Tues.	Wed.	Thurs.	Fri.	Sat.	Sun.
6.1	5.0	4.9	4.3	2.9	1.5	3.0

From their remarks it appeared that Saturday was the favorite day because it is filled with free activities (sports, entertainment, etc.), but also because it is lived with the perspective of a day of rest to follow. The fact that Monday is the day least appreciated is not only explained by the activities it entails, which are the same as for other weekdays, but also because it is lived with the perspective of other working days to follow; as the week goes by, the days appear less and less unpleasant following a regular gradient and Friday, the eve of the favorite day, seems as enjoyable as Sunday. It therefore seems that, although our feelings relate primarily to the *present* activity, they are also strongly influenced by the reality of a near future, while the past seems to have no essential effect on our way of viewing the present.

This point of view is confirmed by a study of the attitude of people living a period of their history which has a fixed end. Soldiers doing military service and prisoners whose sentence is drawing to an end wake up every morning and count the days which separate them from freedom, but they pay no heed to the period which has already been served (Farber, 1944).

If we look closely we find that this is a characteristic of all our behavior. It is always determined by the present situation and oriented by a purpose. The past provides lessons but it has no interest in itself for our everyday life. Our present is directed toward a future, whether near or distant, feared or desired. "Our life is essentially oriented towards the future;" this is the conclusion reached also by Minkowski (*Le temps vécu*, 1933, p. 279). This is the most frequent tendency, but not the only one. We shall now show the vital attitudes which give a value to the present, the future, or the past according to the personality of the individual. In our description of these we shall often have recourse to mental pathology, which acts as a magnifying glass, enlarging these problems by exaggerating the attitudes which an acutely observant critic like Poulet discovered in writers who have found a means of expressing their own temporal horizon through their work.

3. Domination by the Present

There are some creatures who live only in the present simply because they are incapable of forming a temporal horizon. This is true of animals, and we have seen that it is also the case for babies; mentally deficient people who have been unable to construct their past or future are also unable to escape from the present. "They see no further than enjoyment of the present; the rest is more or less outside the bounds of their appreciation." (Minkowski, *ibid.*, p. 335.)

There are, however, other creatures who live chiefly in the present because their temporal horizon has *shrunk*. There can be many causes for, and a variety of aspects to, this involution.

According to Minkowski (*ibid.*, pp. 275-276), manics are sick people whose contact with present reality has remained intact, "but it is only an instantaneous contact" which cannot find a place in a temporal horizon. Such people are hypersensitive to the stresses of the external world: "An object on which their glance falls, an inscription, an incidental noise, a word which happens to ring in their ears, are embraced one after the other in their speech . . . they express their perceptions in words and find themselves carried along, without purpose, by the stimulus thus created." (Kraepelin, quoted by Minkowski.) Their life in the present is obviously very poor—rather like that of an animal or a mental defective—for they are a plaything of the immediate present which is for ever changing. The only information they can give us is negative: they show that life in the present cannot be rich and efficient unless it is part of an organization which integrates the lessons of the past and calls on the future to meet the demands of the present. The euphoria of a manic is also bound up with this shrinking of his temporal horizon: neither the weight of the past nor the uncertainty of the future can influence his mood, which depends entirely on the present.

When old age is accompanied by a weakening of the intellect, it blurs in its own way the temporal perspectives which developed with the intelligence. The old man no longer thinks of the future and his recollections of the past are dimmed. "This impotence of anticipation and imperfection of retrospection condition an unconcern which is not indifference but serenity. . . . The possibility of such complete detachment from the past and the future, from people and things, is perhaps only the natural end of the human mentality when the organism is spared by illness and succumbs to the exhaustion of old age." (Minkowski, *ibid.*, pp. 340-341, quoting Courbon, 1927.)

This last case differs from that of the manic in that the present remains consistent and oriented. It is only the long-distance temporal perspectives which have vanished.

Apart from these examples where the ascendancy of the present is due to a sort of inability to concern oneself with temporal perspectives, there are other cases where the temporal horizon shrinks until only the present is left as a result of the individual's defense mechanisms being turned against dangers from the past or the future which seem to threaten his integrity.

Baruk's description of people who have remained affected by deportation and racial persecution during the war shows us the mechanism of this refuge in the present: "Driven from place to place, oppressed, threatened, terrorized, often faced with an apparently closed and hopeless future, these subjects finally got into the habit of not thinking of the future and also of stifling all memories of their past lives. *They now live only in the present* and they have destroyed the continuity of the past into the future. This fixation in the present of their entire psychological orientation has far-reaching consequences. It destroys the impression of the purpose and finality of the personality and also the concept of the very value of this personality." (*La désorganisation de la personnalité*, 1952, p. 13.) Without being actually overcome by mental disorder, all the people who were taken prisoner or deported felt the need to protect themselves more or less consciously against a past full of happy memories, which would have softened their resistance to each moment, and against the thought of a future whose uncertainty would have added to their discouragement. In their case, the pressure comes from external circumstances.

In the case of neurotics, it is often a result of their own conflicts. When this is so, time may be a defense mechanism which permits them to *isolate* their ego from its traumata or its compulsions by creating an interval (Fenichel, *The psycho-analytic theory of neurosis*, 1945, p. 162). This mechanism may act in different ways according to the direction from which the threat comes. The latter may come from the present, in which case the sick person will seek refuge in the past, for instance; it may on the other hand come from the past—this is the case which interests us—and in the case of obsessive neurosis this may lead to refuge in the present. A neurotic affected in this way endeavors to separate a painful or threatening past from the present. He may for instance become excessively punctual in order to avoid the onrush of instincts and overcome the fear of losing his integrity, or in order to keep himself from "evil urges." Another may insist on not wasting any time so that he will not be submerged by his unconscious, and he may find security in this contact with himself and the world of objects (Dooley, 1941). "As long as the timetable functions as the regulator of their activities, they are sure that they are not committing the sins they are unconsciously afraid of; and as long as they know beforehand what they will do afterwards, they are able to overcome their fear that their own excitement may induce them to do things they are afraid of." (Fenichel, *ibid.*, p. 284.)

According to Poulet, Jean-Jacques Rousseau is one of those who take refuge in the present as a defense against a future they fear. His deranged imagination made him think in advance that the future could not be anything but unhappy. "My scared imagination . . . lets me foresee nothing but a cruel future . . . ," he wrote in his *Confessions*. Thus the compulsion toward the future, which was so strong in him, became "a compulsion towards unhappiness," and he sought happiness in the intensity of his feelings and sensations of the moment. "My heart, mindful only of the present, fills all its capacity, all its space with nothing else." (*Confessions*, quoted by Poulet, p. 171.)

His attitude may be compared with that of Benjamin Constant who was also in search of happiness: "It is the reaction of the past and the future on the present which makes for unhappiness. At this moment I do not suffer at all. . . . What I have suffered no longer exists, what I shall suffer does not yet exist, and I torment myself, I break my heart for those two bits of nothingness! . . . How stupid! . . . It is better to make the most of every hour, uncertain as we are of the next." (Quoted by Poulet, pp. 218, 223.) His conclusion is logical, although his anxiety actually prevented him from keeping to it.

Obviously this attitude, which is one variation of the Epicurean *carpe diem*, is the aim of a man who is content to feel himself live, when the present hurts him less than a past laden with suffering or remorse or a future fraught with anguish through its very uncertainty.

In other cases the double time perspective may be obliterated, not by a desire to protect oneself against the future or the past but simply because the present has a special resonance. If Heymans and Le Senne are right, this is even a trait of character: *primarity*, as opposed to *secondarity*, which from our point of view is defined by the impact of impressions on our life. "Primary man lives in the present and is renewed with it: primarity is the Fountain of Youth. The secondary man, on the other hand, deadens the present . . . by a structure which weighs it down, facing the present event with the repercussion of a multitude of past impressions." (Le Senne, *Traité de caractérologie*, 1945, p. 89.) According to the statistical research of Heymans and

Wiersma, the characteristic traits of a primary individual include the following:

He acts with a view to immediate results,
He is immediately reconciled,
He is quickly consoled,
He is always eager for change.

Whereas the secondary man is:

A creature of habit,
Fixed in his ideas,
Attached to old memories,
Constant in his affections,
Far-seeing in his actions.

Thus the primary man appears to be one who is renewed with the changes in his life, and his present decisions and feelings are not weighed down by a past which obsesses or by projects which bind him. For him the present situation is of prime importance, not because he has no temporal perspectives but because these have little effect on him.

Paulhan described beings whom he called "presentists" and who were characterized by the "excessive predominance in their minds of the mental state of the moment" (1925, p. 193). This predominance results "from the weakness, absence, delay or inadequacy of control; the tendencies which should exercise control do not function" (1924, p. 193). In his opinion *presentism* is not opposed to *futurism* or *pastism,* for present representations rising from our past or concerning future projects can force themselves imperatively on the present moment. Such beings have one trait in common: past experience does not control their present reactions. It is likely that many delinquent children, for instance, are governed by motivations which do not permit them to bear the frustration entailed by the slow realization of a project. This is shown by the research of Barndt and Johnson (1955), who found that stories invented by delinquent boys covered a shorter period of time than those of adolescents of the same age, the same intellectual level, and the same socioeconomic status. Levine and Spivack (1959) made a study along the same lines of adolescents lacking in emotional stability and showed a relationship between the extent of the temporal horizon and the capacity for sacrificing an immediate gratification for a more distant goal.

This sort of person who, in his enjoyment of the present moment, forgets "both what preceded it and concern for what is to follow" has been described in a striking portrait by Alfred de Vigny: His spirit is a "perpetual chameleon; it ends by being neither happy nor unhappy. It is a flame which flares up only when others move and, having no life of its own, remains incapable of *being;* it does not deserve to be relied on, any more than a soap bubble wafted hither and thither by the wind, reflecting the color of objects it meets." To this portrait he opposes that of the soul who is "attentive to the three aspects of his existence all at once, to the past, the present and the future, who does not cease to recall what *has been* through his memory, to consider what *is* by applying his judgment to it, and to speculate as to the probabilities which *will be* through his imagination, subjected to the calculations of reasoning and the laws of his will." (*Journal,* quoted by Poulet, pp. 263-264.)

There is no relationship between being thus the prey of the present and being present to a situation which must be encountered. We are spared this confusion thanks to the work of Heymans and Wiersma. They have shown that it is active, as opposed to nonactive, people who are the most present in their work and the least distracted. But this trait has no connection with primarity or, therefore, with the impact of the past and of distant projects on present experience.

To sum up, time perspectives may be absent owing to a congenital or pathological deficiency or they may be banished by those who fear their threat; finally they may be there, but be underestimated because they are submerged by the resonance of present impressions. These are the different ways in which our experience may be limited to present changes.

Whether the individuals we have been discussing are normal (children, primary adults, or old people) or ill (those who are mentally deficient, obsessed, or manic), we cannot deny that they are lacking in one of the riches of mankind. They are the plaything of unceasing change. Through them we can see what true control of time must be: it demands of man a vision which covers all the experience he has acquired in the past and all the plans he can foresee in the future. Is this not to aspire to the highest level of that function of reality so often referred to by Janet in describing mental health?

4. Domination by the Past or the Future

He who limits his temporal horizon and reduces it to the present alone is without doubt neglecting a part of reality. This behavior still demands, however, that he face the present situation. When it seems beyond our strength to face the present, it remains for us to take refuge in situations where we will find the realization of our desires with less effort. This is precisely the kind of escape afforded by the past and the future, where we can live in imagination. In daydreams as in proper dreams, we are delivered from the pressure of the present, and desires tend to be gratified through fantasms (Bergler and Roheim, 1946).

The refuge offered by daydreams is to be seen in fatigue, in psychasthenia, and in mental illnesses in general. They are usually retrospections, although sometimes anticipations, and they grant satisfactions denied by the present. If we agree that what is subject to the changes of reality is temporal, then we can say that these daydreams are intemporal and thereby we agree with Freud and a fair number of psychoanalysts that they are an eruption of the unconscious, which is also intemporal. However this may be, they are an escape from the present reality.

In the cases quoted above there is an *escape* from the present into the past or the future, but there are some sick people who cannot fix themselves in the present, not because they flee it but most often because the domination of the past is too strong. This is the case for overwhelming remorse. It is in fact the *relative weight* of the past, present, and future at a given moment which counts for the individual. Either he flees an intolerable situation, past, present or future, and thus takes refuge in another time, or he is under the domination of the past, the present, or the future to such an extent that the other aspects of reality are veiled.

These phenomena, which we shall study under the magnifying

glass of illness, are experienced by all of us to some extent. When a child finds his work too difficult, he dreams of some past or future satisfaction. Feelings of guilt and previous frustrations distract us from our present tasks as much as preoccupation with the future. On the other hand, awareness of the present always marks a return to equilibrium of the personality. Volmat (*L'art psychopathologique*, 1955, pp. 167-169) noted that the mentally ill rarely paint scenes which bear a relationship to the content of their illness, whereas their conflicts reappear in their work as soon as their condition improves. A number of authors have also noted that, after lobotomy, patients become aware of the present again. Petrie (*Personality and the frontal lobes*, 1952) asked them standardized questions in normal perspective and found that the number of those more satisfied with the present, or no longer so unhappy with the present, increased after the operation. "Thus, after leucotomy we have a picture of an individual who is much more absorbed by, and lives much more happily in, the present than he did before the operation, who tends to leave the past behind him and who, when he does consider the future, is more reassured. This is in contrast with his state of mind before the operation, which tended to be oriented towards the past and was highly dissatisfied with the present." (*Ibid.*, p. 30.) We should note that this result does not contradict another observation often made on people who have had this operation: they show a certain lack of interest in the future which is not a loss of the ability to foresee but a reduction in the capacity to make the effort through which the different temporal perspectives can be envisaged together (Porot, 1947; Jones, 1949). This is indeed a deficiency, but it is their salvation, for it delivers them from the anguish of the past or the future.

Although present difficulties turn us toward the past or the future, we would stress that it is not usually possible to take refuge at the same time in *both* these perspectives. The two are antitheses in significance and are always relative to each other: when the future is closed the past assumes a disproportionate importance, or it is the domination of the past which makes the future vanish from sight.

We shall therefore consider in turn the two directions in which a man can be oriented outside the present.

a. Escape into the Future. Our life in the present is normally oriented toward the future, which gives us a purpose to our actions. The future is, however, of greater or lesser importance according to the link it keeps with our present activity. It can be the goal of our activity but it may be nothing but escapism if it gives rise to an overexclusive anticipation of what is yet to come. This is the attitude depicted by Montaigne: "We are never at home; we are always beyond it. Fear, desire or hope drive us towards the future and deprive us of the feeling and contemplation of what is." (*Essais,* quoted by Poulet, *op. cit.,* p. 3.)

The young Vigny said of himself: "I do nothing, as you have guessed, but dream of a few projects for the future;" this statement, which could be a trite remark from a young man, takes on its full significance when related to the following: "I have always been so afraid of the present and the real in my life . . ." (both passages quoted by Poulet, *ibid.,* pp. 248, 249). Janet would have been delighted with this link established by Vigny between the present and the real. It is indeed significant.

Does this fear endanger mental health? In general, no. Like many children Perrette may forget she is balancing a milk pitcher on her head and dance for joy at the thought of some future happiness;[11] the damage is rarely serious. The castles in the air of our daydreams imply, however, that the present situation does not absorb us and that we are not very sensitive to the lessons of the past, or that we deny them. When extreme, they are a pathological symptom. Nina, whose story is told by Pichon (1931), gives us an example of a neurosis manifested by escape into the future. Her childhood with a brutal father was unhappy and brought her to prostitution. This past she rejected. When she married she lied to her husband about her parents and suppressed the memories of her childhood. She did not find happiness in marriage. She "could not accept the burden which was forced on her by reality and which hurt her." Her neurosis was manifested by frigidity, but Pichon also points out, and this is what interests us, "that she becomes a prey to so-called

[11](Translator's note.) This is an allusion to La Fontaine's fable, "La laitière et le pot au lait."

'advanced ideas' and frequents the haunts of pacifists, interna-
tionalists, feminists, naturists. . . . She suppresses the collective
past of the society to which she belongs as well as her own
individual past; nothing must survive from the traditions and
customs of past generations, hence her appetite for everything
which smacks of novelty or revolution."

The wish for change is no doubt always the result of dissatisfaction with the present, but it also arises from a feeling that
the future may hold something different from the past. There is
nothing unhealthy in this; we only become unbalanced when
we no longer act with a view to realizing this future in reality
but take refuge in a dreamed or even lived fantasy.

This attitude, however, never points to a condition as serious
as that which occurs when the future seems blocked. Escape into
the future still shows that "psychological strength" which
Eysenck considers to be the conative component of personality.

b. Return to the Past. The past is inherent in the least of our
actions. Our present activity continuously takes into account all
the experience with which we are enriched. But the role of the
past, like that of the future, varies according to the value we put
on it. We may simply use it to realize a future which will be a
new conquest of being; on the other hand, we may refer to it as
a norm. This does not mean that the future immediately closes
up, but it no longer determines the present as a final cause. Instead the present is determined by the past. The drama of
Racine gives us an excellent example of this kind of attitude. It
"is shown as the intrusion of a fatal past, a determining past, a
past which is the efficient cause in a present that seeks desperately
to become independent of it." Racine's works are tragedies of
fidelity. "Fidelity to hatred as in *La Thébaïde*, to love as in
Andromaque, to tradition as in *Bérénice*, to blood as in
Phèdre. . . ." (Poulet, *op. cit.*, pp. 106-107.)

This fidelity implies that the present and the perspectives of
the future are continuously resituated by reference to the past.
Religious ideas, as in the case of Racine, or philosophical notions
may play their part. The social background also often has some
influence. An individual belonging to the upper classes, who
"comes of a good family," is taught from childhood to think of

himself as one of a line. His main care is to respect traditions and keep to his station in life.[12]

Education can generally achieve the same result by creating a strong superego which evaluates the injunctions of the mother and father; temperament also plays a part and *secondarity* is created by the impact of past events on the future.

Often, however, the human being is thrown back on the past because the future seems to be closed for him. There are many causes for this: age obviously, but also illness and personal or social failures. The future no longer seems capable of creating a new situation which could liberate him from fatality. Everything is determined by the past and he is dogged by despair.

In depressive cases, these perspectives can become tragic. As Straus saw it (1928), the more closed the future, the more they feel tied to the past. The future may also often seem negative because it harbors a threat. Minkowski (*op. cit.*, p. 174) has described his observations of a patient who was expecting a horrible punishment for his "crimes." The arms and legs would be cut off all the members of his family and he would suffer the same fate; he would be mutilated in the most terrible fashion and be forced to live with wild animals in a cage or with rats in a sewer. With this prospect "he feels the days pass one by one, uniform and monotonous; he feels time flow by and he complains: 'Another day gone.' No action, no desire takes shape which, emanating from the present, would fly towards the future, above the succession of grey, identical days."

In other cases it is the fear of growing old which gives concrete form to the threat of the future. An example of this is given by Christine, whom Kloos describes (1938). At the age of 30 she complained of living as if the future did not exist. She could not think of it; she felt empty and lifeless. But she could remember the past, though all seemed empty, cold, and colorless.

---

[12] This portrait is even more striking when opposed to that of individuals from other classes. The middle-class man is oriented chiefly toward the future. When he is young he hears repeatedly: "Think of your future," whereas the son of a good family is asked again and again: "What would your grandmother say?" On the other hand, members of the working classes have neither past nor future. What matters to them is to live in the present without concern for a past which inspires no gratitude or a future which is too uncertain (Leshan, 1952).

People suffering from this disorder know that there is a future, as Straus, Kloos, and others have stressed, but it represents nothing for them. It no longer plays any part in their lives except that of throwing them back on the past and often of giving them a heightened consciousness of day to day change, precisely because this has no other sense for them now except that of pure change, like the steps of a man marking time. These are the words of a melancholic: "Now, while I am talking to you, I think as every word is spoken 'past, past, past.' . . . When others talk . . . I cannot understand how they can speak so calmly, without saying to themselves all the time: now I am talking, that will last so long and then I shall do this and that, and that will last sixty years, then I shall die and others will come after me, and still others, they will live as long as I, they will eat and sleep like me and this will go on with no meaning for thousands and thousands of years." (Gebsattel, quoted by Minkowski, 1933, pp. 280-281.)

Does refuge in the past give them a reason for living? Definitely not in pathological cases; the disorganization resulting from the illness does not allow it. There is, however, one famous example of an individual who found his salvation in his past. Proust, seeking for things past, is doubtless a prey to anguish, "the anguish of a being who finds nothing to justify his existence, who is incapable of discovering a *raison d'être* and therefore also incapable of finding anything to guarantee the continuation of his existence; he feels at the same time terror of the future which changes him, contempt of the present which avows its inability to keep him as he is, and the need to flee from this terrible situation whatever the cost by rediscovering in the past the foundation of that being which he is and yet *no longer is.*" (Poulet, *op. cit.*, p. 371.) But Proust was able to give this anguish some *value* in so far as this quest for the past became the *purpose* of his life.

5. Refuge in the Intemporal

This is a survey of temporal attitudes that would be incomplete without any illusion to those particular individuals for whom all temporal perspectives seem to disappear because such people are unable to locate themselves in change. If they are still capable of noticing the succession of events which besiege them, they do not feel their dynamism because they cannot integrate them (Baruk, *op. cit.*, p. 34). At the same time everything seems to come to a standstill within them. They live neither in the past nor the future but in the intemporal or, if this seems ambiguous, in a static present, which ignores what is as much as what was or shall be. Their example shows clearly that our obscure feeling of time and the perspectives it gives us are the result of an assimilation of the changes among which we live.

We can experience this ourselves when we try to retreat from the outside world through reflection or contemplation. We know that schizothymic temperaments have a predisposition for this. But in normal beings this "autism" is rich and reversible. In schizophrenics, on the other hand, "the vital contact with reality" is lost, as Minkowski concludes in his analysis (*op. cit.*, p. 266), and their experience has the poverty of disintegration.

Vinchon (1920) also noted this transformation of the present into an eternity, but it is Minkowski who has given us the most striking examples. To describe this disorder let us quote some of his own observations or his accounts of those of Fischer (1929). "I have a tendency for rest and immobility. I also tend to immobilize the life around me. . . . The past is a precipice, the future is a mountain. . . . To go round in a circle so as not to draw away from the foundation, so as not to be uprooted, that is what I want." (Minkowski, *op. cit.*, pp. 261-262.) "I just can't seem to grasp the fact that time passes and the hands of the clock go round. Sometimes, outside in the garden when they run quickly up and down and round about, or the leaves whirl in the wind, I wish I could live again as before and be able to *run with them* within me, so that time would pass again. But there I stop and I do not care . . . I just bump into time." (Fischer, quoted by Minkowski, *ibid.*, p. 268.)

"Thought stood still, yes everything stood still, as if time had ceased to exist. I seemed to myself to be an intemporal being, perfectly clear and limpid in the relations of the soul, as if I could see my own depths. Like a mathematical formula. . . . At the same time I heard in the distance a silent music. . . . All this in an incessant and continuous flow of movement which contrasted most vividly with my own state of mind. . . . I was as if cut off from my own past. As if it had never been, so near it was to shadows. . . . As if life were only now beginning." (Minkowski, *ibid.*, p. 269.)

The words of these subjects are a symbolical transcription of the disturbances in their experience of time. These have an appreciable effect on their behavior. Minkowski describes this as *"actions with no tomorrow, petrified actions, short-circuit actions,* actions with no purpose." (*Ibid.*, p. 264.)

They are incapable of ensuring the synchronism in their lives of the changes which take place within them and those that occur outside; thus their behavior takes refuge in immobility, which accords well with the domination of space even in their very thoughts (geometrism, reification).[13]

The same cause may produce the reverse effect in other people. In an attempt to resume contact with the changing reality which escapes them, they pack time full with their projects or actions, so as not to waste one second. They have a "tendency to fill time to overflowing with ideas or actions established in advance." (Minkowski, *ibid.*, p. 264.) Sometimes they are obsessed by watches or clocks, clinging to them in the belief that if they let them go, time will vanish (Fischer, 1929, 1930).

These analyses were accompanied by investigations of another kind. Instead of questioning schizophrenics on their impressions of actual experiences, an attempt was made to measure the extent of their temporal horizon by studying their imagination. A group of such patients who were hospital cases, but fairly cooperative, were asked to enumerate 10 events which would

[13] Bonaparte (1939) tried to explain the temporal disorders of schizophrenics by a destruction of the "dykes between the preconscious and the conscious" of great enough extent "to allow a flow of intemporality to rise from the depths of the unconscious, sufficient to submerge the sense of duration or time almost completely." (p. 78.) We consider that this hypothesis complements Minkowski's interpretation. In the case of the schizophrenic, the break between him and his surroundings makes him a prey to his internal fantasms, which are indeed intemporal since they are not subject to the law of change which governs the world and every thought which takes shape within it. It is perfectly accurate to say that a sense of time can only exist where there is submission to reality.

take place during the remainder of their lives, giving the date for each one. The average extension of their anticipation was 12 years, whereas that of a control group of patients hospitalized for illnesses other than psychiatric disorders (of the same age and the same mental level) was 36 years. Even more informative is the test where they are asked to invent stories of the following type: "When Bill wakes up he begins to think of the future; he hopes . . ." The time perspective of the stories made up by the schizophrenics averages 9 months, that of the control group 4 years (Wallace, 1956).

By considering these extreme distortions of the temporal horizon, we have been able to obtain a better understanding of how much perspectives vary from one individual to another. They are determined by numerous factors, of which we have stressed the most important: age, education, social position, and also temperament and mental structure. These factors have their effect on the individual all the time. They each have their own part to play and they shape our attitude to the requirements of time. But other, truly psychological factors also have some influence; they are responsible for the dynamic processes through which we ensure the integrity of our ego. We value the situations which can procure us the most satisfaction or promise us the greatest security. From this point of view the past, the present, and the future are not of the same importance. Our analyses have shown that the normal attitude of man is oriented toward the future and, even when pushed to the extreme, it does not easily become pathological. Its logic always implies a creative effort. Orientation toward the present is also a special case: it requires that we always take the facts of a situation into account and it is therefore essentially attention to reality. On the other hand, refuge in the past or escape into the intemporal are easy attitudes because they refuse to face reality. They are the manifestation of great psychological weakness. These perspectives are of course never exclusive of each other, but we can see the order of their importance for which the criteria are the equilibrium and efficiency of human behavior.

||||||||||||||||||||||||||||||||||||||||||||||||||||||||||||||||||||

THE ESTIMATION OF TIME

EVERY CHANGE INCITES US TO ACT, AS LONG AS OUR REACTIONS ARE determined by the present situation alone, duration, that is first and foremost the interval separating one change from the next, is not a variable of our action. The only exception is perceived duration, for this is embraced in the psychological present (see Chapter 3).

Duration only becomes a psychological reality when the present action does not bring immediate satisfaction. In this case the present phase of change (whether undergone or caused by us) merely announces the approach of another which will correspond to our present expectation. Thus duration has meaning for a dog which hears a sound that is the signal for food to arrive in a few more minutes, or for the rat which must await the right moment to pass from one grid to another, avoid electric shocks, and reach its reward (Chapter 2).

Human beings become conscious of duration in the same conditions. As long as we are concerned only with present situations, there is nothing but "nows" without duration. It happens fairly often that we live for a few minutes, and sometimes even for a few hours, without being aware of duration and without thinking that time is passing, usually until the demands of society force us to replace ourselves in time. We then *know* that time has passed, but we have not actually experienced it.[1]

We must make a distinction between our experience of duration and the knowledge of duration which underlies the development of our temporal horizon. I can *recall* past events, *know* how much time separates them from the present day, and yet not actually experience this duration unless I become aware of it, for instance through regret that those days are gone and a desire to relive them. In the same way I can imagine the future, but as long as I do not long for it or fear it I have no experience of duration. If I read a science fiction novel, I am aware of the difference between the present state of the world and that of the future, but the duration separating these two states is not experienced. In my actions I can take into account what has already happened and foresee what I am going to do or say next, but as long as these thoughts are part of my present action they do not entail consciousness of duration.

The case is quite different when, obliged to perform a monotonous task, I think of the joy I should have in ridding myself of it and doing something else.

Duration is experienced whenever the present situation refers us to another situation in the past or the future. This implies that the present does not satisfy us. Time presents a barrier to our desires. In becoming conscious of duration, we also become conscious of resistance. The word *duration* comes from the Latin *durus*, meaning hard; this double meaning is still apparent in the verb *to endure*. This resistance is manifested in the form of an emotional state which corresponds to our assessment of the value of the obstacle. This consciousness is always expressed as a quality. We become aware of time when it seems to us to be short, or more often long. We think we are apprehending a reality

---

[1] "We cannot say that the experience of time is a primitive experience. It is that of the present, it is the experience of Being; and time is merely an order which we introduce among the modalities of Being. But it is an experience which is in itself derived; it is the product of reflection. It is always slipping away from us and having to be restored when we learn to discern, in the being which we thought established, transient modalities, that which we lack or that which escapes us. . . . When time is filled by our activity we do not leave the present." (Lavelle, *Du temps et de l'éternité*, 1945, pp. 235-236.)

when actually we have only become aware of our own reaction. We therefore agree with Janet for whom duration is primarily a *feeling*, since it is "a reaction to the execution of an action." (*L'évolution de la mémoire et de la notion de temps*, 1928, p. 53.)

Awareness of duration may also arise from consciousness of the changes which took place in it. In fact, the more numerous the changes between us and a moment in the past or the future, the longer the interval thus formed seems to be.

This constitutes the *direct* appreciation of duration, not to be confused with evaluation through actual measurement. The latter is always *indirect* whether the measurement is made by means of natural or man-made clocks or simply by considering the quantity of work produced (ground covered, number of pages read, items manufactured, etc.). When related to these measurements, the number of changes apprehended permits us to appreciate the duration by intuition.

The intuitive nature of these estimations explains why they are never very accurate. However, they play an important part in our lives. We are led by practical necessity to measure time every day. Proper means of measurement are not always available; a baby has to do without them more or less completely. Thus we fall back on criteria—which, however imperfect, are part of our most intimate experience—consisting of feelings of time and awareness of the changes we experience.

Through these two types of information we shall see how time is evaluated.

I *Feelings of Time*

"What's the time?" That is the question we most often ask spontaneously. Usually it merely indicates the necessity of synchronizing our activity with that of our fellows, whether for food, work, or entertainment. It does not imply in itself any consciousness of the duration between two events; we are, however, in the habit of using the answer to this question as a useful guide for the *calculation* of durations.

On the other hand, this same question "What's the time?" may betray a desire or fear of seeing the present period end, and of finding oneself in another period. In this case the question is prompted by awareness of duration.

The case in which our consciousness of duration is most manifest is that of *expectation*. There is expectation when circumstances impose a delay between the first awareness of a need and its fulfillment. But expectant behavior is not induced by any delay. To expect a dear friend to arrive next week and to be on the platform waiting for the train are two different things. In the second case, the expectation is really a specific type of behavior which is well defined by Janet (*ibid.*, 1928, p. 141): "Waiting is an active regulation of the action which comes between two stimuli, one preparing and the other releasing, and which keeps our activity between the two at the phase of preparation." This active regulation consists in trying as far as possible to suppress the expectancy. A young child is not capable of waiting; this gives rise to impatience and even tantrums. He only learns to bear waiting as his emotional stability develops (Fraisse and Orsini, 1955). Even in an adult there are often traces of these anticipations of what is to happen (you go to the window, you imagine what will happen, what you will say, etc.). By learning to defer our reactions, or to *endure* delay, we become conscious of the interval which separates us from the awaited event. According to many authors this is actually the original experience. Time is "fundamentally nothing but the conscious interval between a need and its fulfillment." (Guyau, *La genèse de l'idée de temps*, 2nd ed., 1902, p. 34.) This origin has been rediscovered and appropriated by the psychoanalysts. Some connect the experience with the oral phase and some with the anal phase, but they all confirm that immediate emotional reactions to frustration gradually give way to more conceptualized anticipations where a distinction is seen between the present lack and the future gratification (Wallace and Rabin, 1960).

Instead of being the interval between the awakening of a desire and its gratification, time may be the obstacle to be overcome in order to continue a task which has been undertaken, when the initial impulse is exhausted. If there is still an element of waiting, it is the expectation of finishing the job. Despite the similarity, we should make a clearer distinction between this behavior, which Janet called the effort of continuity, and true expectation. The time to be overcome is that of the duration of the action which must be carried out in order to attain an objective defined by social obligations: to finish one's meal, homework, or working day. The difference between the present result and that which must be realized gives rise to awareness of duration: a child would express this consciousness by such words as "I'm bored" or "it takes so long." The Germans use the word *Langeweile* to express the feeling of boredom caused by a situation from which we cannot escape; the word means *a long time*. "In its purest form," said Lavelle, "the consciousness of time is boredom; that is, the consciousness of an interval which nothing crosses and which nothing can fill." (*Du temps et de l'éternité*, 1945, p. 236.)

Janet even went so far as to say "that the beginning of the duration, the first act which is carried out in relation to the duration, is the effort of continuity" (*op. cit.*, p. 55). Genetically, however, expectant behavior comes before the effort of continuity. The latter requires a conformity with social norms which appears after the duality of the desire and its satisfaction. It is indicative of a self-control even greater than the delayed reaction.

However this may be, waiting and the effort of continuity are the two chief situations in which consciousness of duration spontaneously appears. In both cases it is the result of dissatisfaction. The most primitive feeling of duration, therefore, arises from a *frustration of temporal origin:* on the one hand the present moment does not afford us the satisfaction of our desires and on the other it refers us to a future hope (end of waiting or of the action begun). As long as this frustration weighs on us it is manifested, among other things, by a consciousness of the obstacle, that is of the time interval. Hence this unexpected conclusion: when time becomes a conscious reality, it appears *too long*. "We only find *length* in time when we find it *too long*." (Bachelard, *La dialectique de la durée*, 1936, p. 48.)

We shall have more to say about this paradox, for it is at the root of the problem of the estimation of duration. As Bachelard says, since the consciousness of time awakens when time resists me, I am apt to overestimate it. The simple fact that frustration attracts my attention to the time interval is enough to cause overestimation of its duration. Wundt (1886) analyzed this problem which Katz (1906) expresses so well: "Every time we turn our attention to the course of time, it seems to grow longer." We shall see why this is so in the second part of this chapter. In expectation and in the effort of continuity, the two phenomena of affective overestimation of the obstacle and heightened attention to the changes which separate us from the end, reinforce each other.

But are there not cases where time seems to be too short? According to our first analyses, it may seem that time appears too short when we wait fearfully for what is to come: e.g., the extraction of a tooth or a separation. In this case time is not a distance to be covered as quickly as possible but an interval to be kept. This situation is not, however, a symmetrical opposite of the other. If we take a closer look, we see that the consciousness of time has the same origin in the expectation of a feared event as in that of a future satisfaction: we become conscious of an interval between the present moment and a certain moment in the future. The fear of seeing this interval over introduces a state of uneasiness which allies it to the dissatisfaction observed in the first case. Our attention is drawn to this, and thus we are back at the same conclusion, that time always seems long if we fix our attention on it.

The complexity of the impressions experienced is even more apparent if we differentiate between two types of fearful expectation: that where we fear the end of a pleasant situation and that where we fear a future event. In the first case, time appears long but we wish it were even longer. If we accompany someone we love to the station, the minutes separating us from the departure of the train seem long (attention to each word, each gesture, to everything which seems to speak of the parting) but at the same time we have the feeling that they are too short owing to our emotional attitude; we should like to postpone the inevitable. "I say to this night: last longer," sighed Lamartine, but that night must have seemed long to him, even in his happiness, for he was too conscious of the passing of time. As we shall soon see, it would only have seemed really short to him if he had not been preoccupied with time, or with the afterward. Binet (1903) describes the case of a sick woman who, when she did not sleep enough at night, thought she was obliged to stay in bed the next day. During her insomnia at night she found the time too short because she was always afraid of not "using up" the essential hours of rest before morning.

When we are expecting an unpleasant event, the time also seems longer to us because we turn our attention to it; this impression is heightened by an effect of contrast: basically we wish it were shorter. We actually wish that what we fear would happen as soon as possible to end the tension which becomes stronger and more unbearable as the fatal moment draws nearer. In other words we "wish it were over." This attitude may seem paradoxical but it is confirmed by an experiment carried out by Falk and Bindra (1954). They asked their subjects to produce a duration of 15 seconds several times by pressing a button for this length of time. At the end of their estimated duration, the first group of subjects heard a sound and the second received an electric shock. The results show that the second group produced shorter durations (in other words they overestimated the duration produced in relation to the first group, whereas we might have expected the fear of the shock to make them put off this unpleasant end).

In cases of waiting, therefore, an immediate feeling of time being too short is never found. Time always appears long, but this feeling may be outweighed by desires to see the time of waiting continue or end which are not related to the actual awareness of duration.

This analysis of the situations in which the feeling of a temporal reality can arise may be confirmed by a study of the cases in which we do not have the impression that time has passed. We often have this feeling. We *know* that the hands of the clock have moved but we have no consciousness of it. When does this happen? We can formulate a rule as follows: we are not spontaneously aware of duration when we give our whole attention to the present situation, that is when we are not made to turn to any other time of action through our needs or through social necessity. In other words, we are not conscious of time when we are fully satisfied with the present situation.

Young children give us an example of a life almost always absorbed by what is happening. Their changes in behavior are in some way synchronized with the imperative demands of their surroundings. We have a similar experience when our day is very busy and involves a variety of obligations which have to be met without our having the time to think or to wish for something different. "We continue to live in time for anyone who observes us from outside and is himself conscious of the interval; but where this consciousness ends, when we cannot compare the idea of what is with the idea of what was or shall be, time itself disappears; this is probably what happens in those perfectly distracted forms of existence which are, as it were, beneath temporality." (Lavelle, *op. cit.*, p. 166.) But as Lavelle, who is extremely observant, notes immediately, these lives are only distracted from the point of view of the objects, for in the case of the subject there is always an extreme concentration of attention which explains the harmony between his own changes and those of the world.

Whether this concentration is due to the power of the situations themselves or to the force of a motivation which is fully gratified in the present moment, it produces the same effects. A child at play, a lover carried away by passion, a writer at his work, all are unconscious of time for long moments. This fact has often been observed but not always interpreted properly, through a failure to observe properly our methods of adaptation to duration. Why, for instance, is a writer unconscious of time? We assume that he is entirely absorbed by his work, which means that while he is writing he feels no wish to do anything else (to eat, go shopping) or even simply to stop work because he is tired.[2] But this exclusive interest does not prevent him from being conscious of changes which take place, be it only the number of pages he has written. These changes will provide him, if

---

[2] "A hundred thousand years of meditation like a hundred thousand years of sleep would only have lasted one moment for us, were it not for the fatigue which informs us of the approximate length of our concentration." (Diderot, quoted by Poulet, *Le temps humain,* 1953, p. 201.)

need be, with signs to help him estimate the duration which has elapsed; on their own they do not give rise to a feeling of time, for this depends entirely on the relationship between the subject and his activity.

If we say, as we often read in books, that time *flies* when we are happy, we make a true but inadequate analysis. Satisfaction and unconsciousness of duration are two concomitant effects of an activity which is exactly adequate to the present motivation. Inversely, dissatisfaction and a feeling of duration are both consequences of frustration.

Motivation and present activity may be sufficient to each other on very varied levels of activity. This situation is characteristic of the most balanced lives and of the highest forms of manual, intellectual, and social activity. But it is also to be found on low levels of activity; for instance when we substitute a subjective, imaginary reality for an objective situation in reality in the strong sense of the word. This is the case in daydreams. We know that daydreams are in themselves more often than not the fruits of frustration or of an escape from a reality too difficult to face. But when the individual has reached this lower plane of activity, he can find there a realization which is fully satisfying at that very moment. It is an acknowledged fact that time has no duration in daydreams, even less so than in an activity on a higher level where some difficulty may always crop up. The same experience is found every day in the hypnagogic state preceding sleep or waking. Some chance idea fills the whole field of our consciousness and when a clock chimes in the distance we are amazed that it is so late in the night or the morning. We have not been conscious of duration (Thury, 1903).

The possibility of taking refuge in daydreams also explains why monotonous tasks seem to be relatively brief to many people. Monotonous tasks easily cause boredom, the feeling which results from the *noncoincidence of two durations*, that of the work, which is slow and irksome, and that of our mind, which longs to be elsewhere (Pucelle, 1955, p. 20). Boredom is always accompanied by a feeling of the slowness of change and hence of the passing of time. Yet tasks which may be called monotonous only appear so to a fraction of the workers in the industry, in fact to 25 percent according to Viteles (1952). Among the 75 percent who do not suffer from this monotony, some may find sufficient satisfaction in the tasks, but it seems more or less certain that others adapt to the monotony by taking refuge from it; this is fairly easy as the tasks require only automatic movements (Lossagk, 1930).[3]

Along the same lines, it is possible to understand the feelings of duration which are present or absent in certain cases of mental illness. Janet has described those mental defectives and demented patients in asylums who go for days without doing anything but who do not seem to be bored and in whom there is nothing to show that they have any feeling of duration. He also expresses surprise at asthenics who can stay in bed without seeing anyone for long stretches and do not feel boredom or find that time passes slowly. "These people have lost their feelings; they love nothing, they hate nothing; objects are indifferent to them, in fact they even call them unreal." One of these patients said: "The objects around me are unreal," but she added: "It is very funny how the days fly by now." What does that mean? "Well, I see that it is evening and I see that it is such and such a time by the clock and each time it surprises me because no time has gone by since it was morning." (Janet, *op. cit.*, p. 50.) The words of this patient recall our impressions at the moment of falling asleep. Our affectivity is much reduced and what Janet calls *feelings of emptiness* predominate. Desiring nothing, asthenics cannot suffer from frustration, and particularly not from temporal frustration; no feeling of duration, therefore, has reason to occur.

The same interpretation may be applied to the temporal dis-

---

[3] Our analysis of the feeling of time is confirmed by investigations concerning workers who suffer from boredom in monotonous jobs in industry. In their case, the more bored they are, the longer time seems (Burton, 1943). The most intelligent individuals suffer the most; probably they cannot find satisfaction in routine tasks (Viteles, 1952). Boredom also hits active workers who do not find the tasks demanding enough, or again those who find life dissatisfying in general and who have a tendency toward anxiety and restlessness. The last-mentioned are predisposed to be discontented with monotonous tasks and probably also with any other regular job (Smith, 1955).

turbances of schizophrenics. We saw in the preceding chapter, when discussing the problem of their temporal horizon, that they seem to live in the intemporal. They complain in particular of having lost the *feeling of time.*

"For me time is something very vague. I know very well how to tell the time; I know for instance when it is noon, etc., but I do not have the notion of time. My mind wanders elsewhere. I never know how long I have been doing anything. . . ." (Halberstadt, 1922.)

"My watch works exactly as before. But now I don't want to look at it, it makes me sad. I just can't seem to grasp the fact that time passes and the hands go round."

"Thought stood still, yes everything stood still, as if time had ceased to exist. I seemed to myself to be an intemporal being." (Fischer, quoted by Minkowski, *Le temps vécu,* pp. 268-269.)

Minkowski's interpretation, according to which their vital dynamism is affected, explains perfectly the words quoted above. How could these people feel that time presents an obstacle, since they are overcome with a feeling of immobility? They only know autistic activity in which the realization of their desires is unlimited, and where they therefore do not encounter time (Vinchon, 1920; Minkowski, *ibid.,* pp. 265-266).

By studying these patients we see that they continue to adapt to periodic change, that they perceive time accurately (Fraisse, 1952), that they are even capable of making relatively correct temporal estimations (Clausen, 1950). It has been noted, of course, that they have some difficulty in putting several events in order, especially when these are not contiguous in time, such as Tuesday, Friday, Saturday (de la Garza and Worchel, 1956), but what seems in them to be selectively affected is the feeling of time (Zeitgefühl) and not their physiological clock (Zeitsinn) or their notion of time (Horanyi-Hechst, 1943).

To sum up, a study of the various circumstances in which feelings of time may be manifested shows that the latter have their origin in the consciousness of frustration caused by time. Time either imposes a delay on the satisfaction of our present desires or it obliges us to foresee the end of our present happiness. The feeling of duration thus arises from a comparison of what is with what will be, i.e., from awareness of the interval separating the two events. On the other hand, when we find—on a high or a low level of activity—the integral realization of our desires in the present moment, we have no feeling of the duration of time. In everyday life these moments are, of course, of limited duration, for fatigue or even a slight feeling of tiredness will speedily conflict with the motivation controlling our activity; even more frequently, the necessity of keeping the social background in mind prevents us from becoming entirely absorbed in our work, however easy and interesting it may be in itself. At special times, however, we can have the impression of becoming independent of time because we are living body and soul in the rhythm of present changes.

II *The Appreciation of Duration*

Our everyday experience shows that we are not very exact in our appreciation of duration. We find it difficult to estimate how long we have been eating a meal, working, walking, reading, etc. We make glaring mistakes when we have no clock or watch.

Nevertheless, we never stop assessing the duration of our activities. Even when we can measure them objectively we still like to compare our intuitive estimation with the measurement.

On what indications do we found our time judgments? Before considering this basic problem, we must discuss the forms taken by these judgments and their respective value. This methodological introduction will help us to give the right interpretation to the research we draw upon.

1. The Modalities of the Estimation of Duration

Our judgments take three forms:

1. We are often content to express our estimations by absolute judgments: "it takes so long," "that didn't take long." Like all absolute judgments, they are actually nothing else than implicit comparisons. "It takes so long" expresses our appreciation of the duration experienced, in relation to some standard. This standard is provided by our anticipation of the probable duration of the

action when we begin it,[4] which depends on our habits (average duration of a meal) or on our desire to finish the action quickly or not. Obviously such estimations are very much influenced by the feelings of time which may be occasioned by the action. It is evident that this method of appreciation is genetically the first to appear. A child finds that walking or eating his food takes a long time.

The comparison is sometimes explicit: "The journey seemed longer than yesterday." This case is the same as the first except that it explicitly involves a previous appreciation of time, and we shall see (p. 233) that our retrospective time judgments are sometimes different from our immediate judgments.

2. We sometimes try to evaluate durations quantitatively by using conventional units of time, minutes and hours (obviously in the absence of a clock). This method of evaluation is only possible after long training which is provided by the repeated use of watches and clocks, but it never becomes perfect because these units have no tangible reality and do not give rise to images. I can form a representation of a yard, but not of a minute; I can only try to reproduce a similar duration. Therefore, in estimating time we try more or less clumsily to put a subjective appreciation into concrete form.[5]

This method of estimation by a translation into units of time is the one used most readily, for it gives information which can be compared with that provided by our watch. It is also the method most often used in experimental studies because it is the most practical.

3. We can also translate our appreciation of time by a method of reproduction. The subject who has had an impression of duration while performing a task is asked to produce an equivalent duration either by working again for the same length of time or

---

[4] We use *action* in the widest sense of the word. Action is what we do, and we are acting from morning to night. Looking, hearing, waiting, are actions just as much as writing, walking, or making something.

[5] The use of exact conventional units represents a great improvement on the use of units drawn from our tangible experience, such as those used by certain tribes of uncivilized natives. In Madagascar, for instance, the natives speak of the duration of "cooking rice," meaning about half an hour (Klineberg, 1957).

by defining by two signals, one at the beginning and one at the end, a period of equivalent duration. This method has the advantage of not involving abstract units. We obviously hardly ever use it in everday life where it would have no purpose; on the other hand it is a very successful method for use in experiments.

4. We can also use the method of production in research. This consists in asking a subject to do something for a duration expressed in units of time, for instance to write for one minute.

These different methods do not produce the same results. This can be checked by using—with caution—the technique of correlations. Clausen (1950) found that the results he obtained by the method of reproduction and by that of estimation had no correlation. In a very systematic study using five durations from 21 to 45 seconds, we found the following correlations for 22 subjects (Fraisse and colleagues, 1962):

Between reproduction and estimation $r = +.17$
Between reproduction and production $r = +.10$
Between estimation and production $r = -.37$

The only significant correlation is that between estimation and production, both of which involve subjective appreciations of time units. We should not be misled by the negative nature of the correlation. A shorter production corresponds to an overestimation in time judgment, and the negative is due entirely to the fact that the correlation was calculated from the actual results obtained and not from percentages of over- and underestimation. Each subject has, however, his own scale of measurements and his own consistency when using each of the methods of appreciation (Myers, 1916; Korngold, 1937; Harton, 1939; de Rizende, 1950; Eson and Kafka, 1952).

None of these methods of appreciation is by any means perfect. Considerable mistakes are made, even for short durations (of a few seconds or a few minutes). Bourdon (1907), who used the reproduction method, estimated that these mistakes can reach from 20 to 25 percent for durations from 9 to 25 seconds and 33 percent for 76 seconds. Woodrow (1930) used the same method but had the same interval reproduced 50 times in succession and found a variability of 17 percent between 6 and 30 seconds. Pumpian-Mindlin (1935), using production, found an average error of 25 percent for durations from 30 seconds to 10 minutes.[6] Gilliland and Humphreys (1943) combined their results for the three methods (reproduction, estimation, and production) and calculated that the percentage of error decreases as the duration increases, from 28 percent for 14 seconds to 18 percent for 117 seconds.

These last two authors also found that the error was less by the method of reproduction than by that of production and also less by the latter than by the estimation method. We found similar results in the research mentioned above (Fraisse and colleagues, 1962). The variability from one subject to another and in one subject (intersubject and intrasubject) is greatest by the estimation method. We give below the results for the standard deviations in the appreciation of durations from 21 to 45 seconds.

STANDARD DEVIATIONS

	Intersubject	Intrasubject
Reproduction	13.8%	20.9%
Estimation	35.1%	28.7%
Production	23.5%	18.8%

The less the units of time are involved in our appreciation, the more accurate we are. The differences observed are far greater in the case of children and mentally ill people, especially when the latter are to a greater or lesser extent mentally deficient. Kohlmann (1950) made his subjects estimate various durations of up to 3 minutes by the methods of estimation and reproduction. The relative errors of 10 normal adults range, according to the subject, from 12 to 28 percent for reproduction and from 30 to 78 percent for estimation. In a group of 12 mentally sick subjects, including cases of cerebral tumors, schizophrenia, and senile

[6] These results all show that Weber's law seems to be applicable to relatively brief durations.

dementia, the errors by the reproduction method ranged from 18 to 108 percent and by the estimation method from 47 to 432 percent. The extreme values, whatever the method, are obviously those given by patients with senile dementia.

The direction of the error varies, as we shall see later, with the nature of the task whose duration is to be judged. It also depends on the experimental conditions: if, during the same experiment, one subject has to evaluate different durations, he will overestimate the short ones and underestimate the long ones as a result of the development of a central tendency (Fraisse, 1948; Gilliland and Humphreys, 1943; Clausen, 1950). This is a general law which also applies to perceived durations (Chapter 5, p. 120).

This discussion of the methods and the variability of results shows that an analysis of the factors and laws governing the appreciation of duration is a very delicate matter.

Work based on the reproduction method is the most reliable but it only permits us to consider short durations. The estimation method makes it possible to compare the apparent durations of several different tasks, since we have seen that each person seems to have a fairly consistent subjective scale. If used alone, however, it does not allow a conclusion as to whether the subject has overestimated or underestimated a duration, that is, to know whether he found the time long or short. If a person judges one hour to be 80 minutes, this does not necessarily mean that he found this hour very long; it all depends on the use he makes of the units of time. On the other hand, there is some sense in the comparison of the estimations of one subject in that these refer on the whole to the same standard. The absolute value of an estimation therefore counts less than its relative value compared with that of another estimation.

2. The Criteria for the Appreciation of Duration

Our appreciations of duration vary a great deal according to the concrete situations. What are, then, the factors which determine these variations?

In estimating duration, we base our judgment on three sorts of indication; we have already discussed certain of these, but it would be useful at this juncture to consider them all in the same perspective: measurement, affective information, and direct information, i.e., information based on the number of changes experienced.

a. Appreciations Based on Measurement. We know from experience that our direct estimations of duration are very inaccurate; we therefore nearly always try to use an instrument for measuring time. Clocks and watches provide us with the ideal means, but obviously the hands of the clock turn at a uniform speed and a calculation of the amount they have moved bears no relationship to our actual experience. This process is similar to spatial measurement, for we only have to note the position of the hands at the beginning and end of the period in order to make the calculation. Of course, the actual measurement of time implies a concept of homogeneous time, but we shall leave this point until later (Chapter 8, p. 268).

When we have no watch, we try to replace it. For long durations we can use natural clocks; for instance, we can estimate the distance the sun or the shadows have moved. We also have at our disposal all the periodic changes, including those of our own organism, which we know take place at certain times.

Another practical method of measuring time is to apply the principle of the clock. This consists in measuring a movement of uniform speed. This particular movement is of course adjusted to the day-night cycle, but any other movement can be used in exactly the same way. Of two distances covered by two objects moving at the same uniform speed, it is simple to infer that movement over the longer distance lasts longer. It is even possible to make the measurement if the unit of movement is based on the movement of a clock. We know, for instance, that if an adult of average strength has covered 3 miles, he has been walking for an hour. Thus we can estimate duration according to the distance covered. The same is true for any work done by a machine or by a man, if the work is of a uniform nature and can be divided into units of quantity. The number of pages written, the length of trench dug, or the number of items manufactured can thus provide us with a basis for the measurement of time.

The less well-defined the units of time, the more difficult it is to determine their number and the more inaccurate our measurement becomes. In this case we prefer to talk of the appreciation or estimation of time; essentially, however, this is still measurement, using more or less explicit calculations, and like any measurement it is an indirect process which has no reference to our actual experience of duration.

b. Affective Judgments. We saw at the beginning of this chapter that our most basic knowledge of duration arises from feelings of time. In their own way these give us the elements of judgment. Either we become aware of duration, in which case time seems too long, or we are not conscious of duration and very little time seems to have passed. This happens every day. If we are waiting for something to happen and we look at our watch, we are always surprised that it is not later. On the other hand if we are spending an enjoyable evening with friends, we are often very surprised by how late it is when we see the time. These judgments based on our feelings of time—or their absence—are not isolated. They color our fundamental time judgments in which, as we shall see, there are other criteria involved. They also have the effect of accentuating these. When a feeling of time arises, our attention turns selectively to the duration and time seems to pass more slowly. "A watched pot never boils," as the saying goes.

c. Direct Judgments of Duration. Let us suppose now that we have no means available of telling the time and time is also not dragging for us. We are none the less conscious that time has passed and we feel capable of judging it.

What does duration consist of? Of successive changes and nothing else. To put it more clearly, psychological duration is composed of psychological changes; that is, of changes which are perceived and therefore have psychological reality. Perception is therefore at the heart of the matter.

Under what conditions is a change perceived? We must answer this question first, for it will throw light on the whole problem.

Of the countless changes which take place around us, we do not notice all to the same degree. The clock behind me is ticking away the seconds, but as a general rule I do not notice every second, unless I happen to be doing something like timing a race. I read a book. Every movement of my eyes entails a change, but I do not perceive it: I hardly even notice turning over the pages.

The changes I experience relate to the contents of the book.

Our perception of change is like all perception. It always has two components: the stimuli and our attitude. Of the multitude of different stimuli which are acting on our receptors all the time, we perceive selectively either the most intense stimuli or those which pertain to our attitude of the moment. A sort of seesaw is thus established between the force of the stimulus and that of my attitude. What I perceive is the result. If I am absorbed by my work, I do not hear—or hardly hear—the children quarrelling in the next room. But when they start to shout, the noise is forced on my attention and becomes the predominant figure while the rest, my train of thought, becomes a vague background.

The selection of perceived changes depends on objective and subjective factors alike. The first relate to the actual nature of the task and the second to the attitude of the subject. This is an abstract distinction; in reality the changes we apprehend depend on both of these. It is, however, possible to a certain extent to decide after a perception on how much of it was due to the stimuli and how much to the subject. For instance, in the Rohrschach test I can distinguish between the ink-blot aspect and my subjective interpretations. When confronted with changes, I can pay attention to the succession of the events which are taking place as it were outside me, or to the effect they have on me. When reading a novel I can judge the number of changes according to the number of pages I turn or I can consider nothing but the adventures I live through with the hero of the book.

This distinction between the objective and subjective elements of change may be compared with that suggested by Straus (1928), based on his pathological studies, between *world time* and *ego time*. According to him, we live simultaneously in two times, one marked by the changes which take place around us, the other being part of our intimate experience and bound up with our personality. Schizophrenics and melancholics are very sensitive to the conflict which may arise between the immobility of their own emotions and thoughts and the changes which they see taking place around them, in which they no longer play an active part. Other ill people may observe a difference in rhythm between their own personal changes and those occurring around them; such is the case of the ill woman described by Kloos (1938) who, during her attacks of melancholia, found that the time by the clock slowed down in relation to her work, which she had the impression of doing very quickly.

We often have similar experiences. These examples are an exaggeration of the differences we frequently find between the speed of changes which actually take place and those we should like to see. They also point to our ability to dissociate the flow of our thoughts and emotions to some extent from that of our perceptions. There is, however, a limit to this dissociation. It cannot be said that *perceived* changes do not also depend on the personality as a whole. The distinction between *ego time* and *world time* is only valid if they are interpreted not as two different things but as two aspects of the same reality which can be dissociated to a greater or lesser degree according to the effect of the attitudes arising from the will or from a disease.

If we accept these premises, we can go on to show that the length of a duration depends on the number of changes we perceive in it.

In so doing we unite ancient and modern viewpoints. When Aristotle found that time was the quantity of movement, it seems to us that he projected a primarily psychological fact into the physical world. Condillac, while concerned chiefly with the origin of the idea of time, thought that his statue would have known no more than an instant if the first odoriferous body to come along had acted upon it consistently for an hour or more; he notes on several occasions that time only consists of the succession and number of impressions sensed by the organ or recalled by the memory. James writes that "it is but dates and events representing time, their abundance signifying its length." (*Psychology, briefer course*, p. 283.) Guyau endeavored to make an exhaustive survey of the factors which have some part in our estimations of time. He found ten, but this number depends entirely on the number or variety of images and everything which accompanies these: emotions, appetites, desires, affections (*La genèse de l'idée de temps,* 2nd ed., 1902, pp. 85-86).

We also intend now to make a systematic study of this question, with the purpose of establishing the following law: any factor which contributes toward an increase or decrease in the number of changes observed has the effect of lengthening or shortening the apparent duration.

1. The Influence of Our Attitude and Motivations

Our attitude can have the effect of increasing or reducing the number of apparent changes. To what extent do these variations affect our estimations of duration? We can determine this by modifying the attitudes while the task remains objectively the same.

We are able to increase the number of changes we perceive by paying attention to the different steps of a task. We have already mentioned the basic law expounded by Katz. The more attention we pay to time, the longer it seems. When we "pay attention to time" we are doing nothing other than paying attention to the various changes which take place. Never does a minute seem so long as when we watch the second hand of a clock move round the 60 divisions on its face. We could give many examples. In most experiments on time judgment, the durations are overestimated when the subject is forced by his instructions to pay attention to the passing of time. If, however, the duration of the same task is estimated several times in succession, this overestimation is found to decrease gradually (Falk and Bindra, 1954). This is a familiar phenomenon; as we get into the habit of walking along a certain street or doing some routine job, it seems to take less time. The explanation is simple: novelty attracts attention and not a single detail escapes us, but as the action becomes automatic we concentrate more on its goal.

We also pay a great deal of attention to changes when we have to make an effort to accomplish a task which is too difficult. But this difficulty is not only inherent in the nature of the task itself; we all know how it is increased by insufficient motivation.

The same effect is noticed if we are forced to wait, if the task is too difficult, or if there is some danger involved. The last factor has been well illustrated by the recent work of Langer, Wapner, and Werner (1961). Sixteen subjects were asked to estimate a duration in the following circumstances: they were placed on a trolley which moved on rails at a constant speed, and instructed to drive it forwards for 5 seconds (production method). They were blindfolded while on the move, but were told beforehand to inspect the route; in one situation there was danger, for the corridor in which the experiment was made ended in a precipice, the stairwell, and in the second situation there was no danger, for the trolley moved in the opposite direction, away from the stairwell. Starting 15 feet from the danger and moving at a speed of 2 miles per hour, the estimation was 3.37 seconds with danger and 4.22 seconds without.

Inversely, everything in our attitude which helps to reduce the number of changes perceived also reduces the apparent duration. The organizing activity of the human mind plays an essential part in this reduction: instead of considering each element of the task in itself we can concentrate on its purpose. This purpose is sometimes imposed by the nature of the task; to arrange figures and to multiply them are two different things. But it may be independent of the task itself and depend directly on the attitude of the subject. This attitude in its turn is a function of the motivations behind it; we work toward a goal if we expect some gratification from it.

It is therefore ultimately our motivation which modifies our subjective evaluation. When this motivation is slight, our attention turns to the various steps of the task; we are also easily distracted by outside incidents or by chance thoughts, or we may concentrate on the effort involved, as mentioned earlier. When the motivation is very strong, on the other hand, we are absorbed by the task itself, which takes on a unity of significance, and we realize ourselves that we are not aware of the passing of time. If a lecture or a discussion is fascinating, it seems far shorter than if it is dull. Workers know well that if they work more actively, or take more interest in what they are doing, the time will pass more quickly (Jahoda, 1941). The same point is shown by the following experiment: The subjects are given similar puzzles (which look easy but are in actual fact insoluble) in two different situations. In the first, the task is presented as a training for the solution of another puzzle later; in the second, the puzzle is presented as the actual aim of the experiment and a time limit is set for its solution. Rosenzweig and Koht (1933), the authors of this experiment, found that 51 subjects out of 89 estimated the time to be longer in the first situation, when the task was presented as a preliminary test and stimulated their interest less than in the second situation. The results would have been more conclusive if the authors had tried to establish what interest the subjects actually had in their task in the two situations.

This criticism is shown to be very relevant by Meade (1960a), who used the same technique and found that the greatest variable was not the level of motivation but the order of the situations. Whether the second situation is that with the high or low level of motivation, the duration being equal, it always appears to be shorter, on an average, than the first situation. If two different groups of subjects are used, the so-called effect of motivation is no longer apparent at all. Perhaps the extent to which the ego is involved in the task does not vary with the instructions received. The nature of the task has a specific effect.

There are, however, other means of varying the motivation. One of these is to make the subject foresee success or failure. Still using the technique of Rosenzweig and Koht, Meade (1960b) found that subjects who were given the puzzle as an intelligence test estimated the duration of the task at 3.4 minutes when he gave them to understand that they would be successful— by saying "good" to them ten times during the 5 minutes which the task actually lasted—and at 5.5 minutes when he gave them no indication of progress. This expectation of success or failure has little effect, however, on subjects who are given the puzzles as an exercise with no particular purpose.

This experiment confirms the results of some older research. The same subjects were asked to do two similar tasks (to learn a mental maze). The experimenter varied the motivation by causing them to foresee success or failure. He warned them that the test was to be completed within a limited time and decided, without telling them, whether they would succeed or fail; he was able to obtain the foreseen result by altering the paths of the maze without their knowledge. If the subject was intended to succeed, he encouraged him during the test by saying that he was on the right track. If he was to fail, on the other hand, the experimenter warned him again and again that he was going wrong and thus made him foresee failure. Fifty-two out of his 57 subjects estimated that the task in which they succeeded was shorter than the other (Harton, 1939b). The author concludes explicitly that stronger motivation leads to better organization of the work, or in other words to greater unity.

Strong motivation may have very different causes. It may result simply from the fact that the task is sufficiently difficult. We are in fact interested in tasks which offer a certain amount of resistance—*without being too difficult*—more than in tasks which are too easy. Harton (1938) showed that if we increase the difficulty of a task, without fundamentally changing its nature, it will seem shorter. One of his experiments involved the comparison of weights. In one case the differences were only just noticeable and the task was therefore difficult; in the other they were obvious. The time was objectively identical but seemed shorter in the first case.

Two examples taken from history seem to confirm this influence of motivation. The first is that of the French miners involved in the disaster at Courrières in 1906; they were trapped in a gallery from which they only managed to escape after three weeks of effort. On emerging they all declared spontaneously that they thought they had been trapped for four or five days. Similarly, at the time of the earthquake at Messina, three brothers were trapped under rubble for 18 days and when they were freed they also thought that only four or five days had elapsed (Ferrari, 1909; Peres, 1909).

The magnitude of these errors can only be explained by the fact that the victims, being buried alive, had absolutely no temporal cues. It seems that they judged the time to be so short because they were obsessed with one idea, the will to survive, and the only thing they could think about was their release. The fact that they were waiting for this to happen should have made the time seem long to them and it is probable that they actually did feel this. But they had absolutely no basis for their estimation of such a long duration. The tension they were under and their weakness must have subdued all the physiological variations which could otherwise have guided them.

2. The Influence of the Nature of the Task

From the psychological point of view, most tasks can be divided into a certain number of parts. The number of these can vary for the same duration. Let us suppose that I have 40 letters to copy. These letters may be independent of each other, in which case I have 40 different symbols to copy. On the other hand they may be grouped into words and constitute only 10 different elements; finally these words may form 1 sentence. The laws of Gestalt psychology also apply to successive changes. What effect do these organizations have on our estimation of time?

This question bears some relationship to the point we have already discussed: the unity of a task cannot exist independently of the subject; and the more unity it has, the more it is likely to seem interesting. Unity reinforces motivation and thus introduces a subjective element. This interaction of objective and subjective elements may make our discussion more difficult but it will not lessen the scope of our conclusions. However complex the interpretation, one fact remains constant: the greater the number of changes observed, the longer the apparent duration.

a. The Influence of the Unity of the Task. This problem is raised in two general studies. The first is that of Axel (1924). He gave 68 male and female students a number of tasks: to estimate the duration of an empty interval, tap with a pencil on a sheet of paper, cross out signs, find analogies, complete a series of figures. The subjects were asked to estimate in seconds the time the tasks took them, these durations varying objectively from 15 to 30 seconds. The average value of overestimation (+) or underestimation (−), calculated from the means for all the subjects and all the durations, was as follows:

Empty time	+1.8 sec.
Tapping	+2.4 sec.
Crossing out	−5.7 sec.
Analogies	−7.6 sec.
Series of figures	−9.2 sec.

Gulliksen (1927) studied the same question with a great number of subjects (326) and a wider range of tasks. These all lasted

200 seconds (there were also some other tasks of different durations to avoid an effect of uniformity, but these will not be discussed here). If the tasks are arranged in order of the decreasing value of the average estimations, this order is as follows:

	Average Estimation (in seconds)	Standard Deviation
Resting and trying to sleep	241.7	107.8
Holding the arms outstretched	228.4	96.2
Listening to a metronome		
66 per minute	223.7	92.4
184 per minute	214.1	85.2
Pressing a point on the skin	210.2	78.4
Reading a passage in a mirror	181.8	77.6
Taking a dictation	174.6	77.4
Doing a division	168.9	70.2

These results were confirmed recently by some important work by Loehlin (1959). He used a large number of subjects, asking them to estimate the duration of 16 different activities each lasting 2 minutes. He found that his activities fell into much the same order.[7]

These experiments show that the apparent duration of the tasks decreases as the activities are less broken up; that is, as the changes are less numerous. This conclusion is particularly evident if we take into account the units of significance which have the effect of reducing the apparent number of changes. While tapping or holding the arms outstretched, we are present to every moment of the duration. In taking dictation, on the other hand, we are writing out sentences or at least groups of words. We also

[7] He also found fairly good correlations between all the estimations; from these he established a common factor. These correlations are explained by individual differences of two kinds: a. The interest of the subject in the proposed activities. There is a correlation of $\rho = .61$ between the mean of the estimations of the boring, as opposed to the interesting, nature of the activity on a scale of 5 points, and the mean of the estimations of the duration. b. The relative scale of verbal estimations of the subjects, the use of the units of time varying greatly from one subject to another.

have a purpose, the faithful reproduction of a text, and this makes the rest seem less important. Therefore the more unity a task has, the less time it seems to take, for the partial changes are no longer in the foreground of our attention. The unity of significance makes the task more interesting, as we have seen, and the subjective and objective elements thus reinforce each other.

Some work of a more analytical nature by Harton (1939a, 1942) has confirmed the results of these experiments. Harton's subjects were asked first to estimate the duration of a task which had particularly marked unity, learning *one* fairly difficult mental maze, and then to estimate the duration of a more disjointed task, learning *several* small mazes of the same type. The total duration of the two tasks was the same, but it was estimated to be 305 seconds in the case of the single maze and 444 seconds in the other case.

Pierre Janet frequently stressed the relationship which exists between the level of behavior, the function of reality, and the unity of the task. Our interpretation of the above results by the unity of the task is also similar to that proposed by Axel (1924) and Dewolfe and Duncan (1959) in terms of the level of behavior. The research carried out by these last-named authors is particularly fruitful. They chose three tasks corresponding to three levels of behavior (resting and doing nothing, printing the alphabet in reverse, solving anagrams); the subjects carried out one of these activities, called the reference, for 26 seconds, and were then instructed to perform another, or comparison task, and stop when they estimated that they had been doing it for the same length of time. This test was done for all the possible combinations of the three activities and the results were found to be very systematic, as can be seen from the following table (mean of the logarithms of the durations):

Reference Task	Comparison Task		
	Rest	Alphabet	Anagrams
Rest	1.42	1.54	1.73
Alphabet	1.25	1.43	1.62
Anagrams	1.27	1.41	1.46

The time judgment varies directly with the level of the comparison task and inversely with the level of the reference task. The comparison between identical tasks (diagonally) is always more or less the same and is not far from the value of the reference (log. 26 = 1.415).

Some further information may also be derived from Axel's results. He questioned his subjects concerning the criteria they had used in evaluating the duration of the various tasks they had performed (they had been told beforehand to use any means of estimation possible, except of course a watch). The following table is very revealing:

	Counting Numbers or Movements of the Body, %	Quantity of work, %	Quantity of work and Necessary Energy, %	Pure Guesses, %
Empty Time	97.1	0	0	2.9
Tapping	61.8	33.8	0	4.4
Crossing Out	11.8	80.9	2.9	4.4
Analogies	0	14.7	77.9	7.4
Series of Figures	0	8.8	80.9	10.3

It is immediately apparent that the subjects did not choose their means of estimation. These were determined by the task. It is impossible to count numbers or respirations at the same time as finding analogies or completing series of figures. The quantity of work has a definite meaning in a regular and uniform task like tapping or crossing out, but is far less easy to determine in work which involves quality. In this case the subjects not only consider the quantity of work but also the quantity of energy expended in carrying out the task, the mental effort involved, etc.

If we compare the use of these means with the results already given, we can see that the time is estimated to be longer the greater the number of changes on which the subjects base their judgment (counting numbers, beats, estimating the number of signs crossed out). On the other hand, time appears shorter in the more complex tasks, when the subjects cannot use such means. In the latter situations the percentage of cases in which they consider their estimation to be a pure guess, i.e., to have no basis, increases considerably.

These results also explain why the time always seems shorter when we are doing anything at all than when we are doing nothing. When we do nothing we do not create a mental void; we either wait for the end of this period, in which case the feeling of time arises and consequently the apparent duration increases, or we observe everything which happens during the period in order to fill it. On the other hand, work of any kind implies a certain purpose, even if a series of disjointed tasks have to be performed to attain it.[8]

b. The Influence of Changes Which Are Only Undergone. Let us consider the reverse of the situation we have just analyzed; when the changes we experience cannot be unified, time always seems to be long. This happens whenever we undergo changes instead of creating them. The best example is that of perception. In this case we apprehend changes without arranging them in large unities because at any given moment we cannot foresee the stimuli which will later occur. A multitude of apparent changes is imposed on us. This is why listening to a text seems longer than copying it (Swift and McGeoch, 1925), and listening or reading seems to go on longer than taking dictation when the objective duration is the same (Yerkes and Urban, 1906; Spencer, 1921).

A wide range of examples could be given. Myers (1916) asked a number of spectators at a basketball match to estimate the time from the beginning of the game until a serious accident, which was objectively 6 minutes, 15 seconds. Eighty percent of the spectators overestimated this duration. The mean of the judgments of a group of 68 men was 10 minutes, 7 seconds and that

[8] From this point of view, the same results are also confirmed by those of Dobson (1954), obtained with 16 subjects and durations of 17 seconds, 38 seconds, and 2 minutes. The subjects had to estimate these durations, during which they had either done nothing or had been placing pegs (Purdue pegboard). The mean of the estimations for the 2-minute period was 210 seconds in the first case and 173.4 seconds in the second. The same results are obtained by the production method (giving a signal when a specified interval of time has passed). The subjects estimate that 2 minutes have passed after 81.7 seconds if they have been doing nothing and after 107.7 seconds if they have been working.

of a group of 32 women, 15 minutes, 54 seconds. These are appreciable overestimations. Musatti (1931) made a film strip lasting 40 seconds; his group of 36 subjects estimated its duration at 2 minutes, 9 seconds. In the course of our own research on remembering films (Fraisse and de Montmollin, 1952), we asked our 115 subjects to judge the duration of the film sequences we had shown them, without having warned them in advance that they would be asked to do this; the question was in any case only secondary in our research. One sequence, which lasted 2 minutes, 47 seconds, was a short dramatic narrative; the average evaluation was 5 minutes, 54 seconds. The other was a news sequence lasting 3 minutes, 14 seconds, and was estimated at 6 minutes, 59 seconds. In both cases the estimated duration was more than double the true duration; this is a far greater overestimation than the usual systematic errors.

These experiments were all carried out on spectators; they are interesting in that the intensity of the motivation produces in these subjects the inverse effect to that which was noted in the case of subjects who had to do something themselves. For a spectator the interests, although real enough, does not give the task unity. The act of perception has its end in itself and not in an objective to be reached or a task to be carried out. It is a pity that Myers' experiment could not be extended to the members of the basketball team themselves. We presume that their estimation of the same period of time would have varied in the opposite direction to that of the spectators.

We could compare this with situations in which we are carried away by a mental "film" summoned up by our imagination when we have no control over it. This is the case in intoxication with hashish or mescalin and in dreams. In all these cases time seems very long because countless images succeed each other with very little connection and we have no possibility of escaping them or referring to other criteria for the estimation of their duration.

The first observations on intoxication of this sort were made by Moreau de Tours (*Du haschisch et de l'aliénation mentale,* 1845). He noted that, under the influence of hashish, one's head feels like a volcano, sensations and feelings follow each other with incomparable rapidity, the flow of ideas seems inexhaustible: he also observed that time seems to drag unbearably, minutes become hours and hours are like days (p. 685). These observations have since been confirmed by many authors, among them some who have been particularly concerned with this aspect of the estimation of time (Pick, 1919; Bromberg, 1934). Favilli (1937) added further details. He found that, after mescalin intoxication, his subjects were under the impression that the period of time involved had been very long but, when asked for a precise estimation, they underestimated it considerably. In his experiments intoxication was not complete, but nevertheless his subjects failed to find any cues to guide their evaluation; they did not trust the changes they had experienced, knowing these to be imaginary, and thus no basis remained for their estimation of a duration which seemed to have no depth.

An effect similar to the well-known influence of hashish and mescalin is produced by a number of products which have a pharmacodynamic effect on the estimation of time. Our knowledge in this field is still very restricted and the results sometimes differ from one author to another. This need not surprise us, however. Ethical considerations and the necessity of obtaining the cooperation of subjects oblige us to use small doses. One product, moreover, does not necessarily have the same effect on all temperaments.

As a general rule, however, it can be said that all drugs which accelerate the vital functions lead to the overestimation of time and those which slow them down have the reverse effect.

Thyroxine (Sterzinger, 1935, 1938), caffeine (Frankenhaeuser, 1959), and metamphetamine (Frankenhaeuser, 1959) have the effect of making us overestimate time while pentobarbital (Frankenhaeuser, 1959) and nitrous oxide (Steinberg, 1955; Frankenhaeuser, 1959) make us underestimate it. The latter effect is also found in subjects living in a rarefied atmosphere (Barach and Kagan, 1949).

It is therefore likely that stimulants induce greater mental activity whereas inhibitors reduce it.[9]

[9] This explanation of pharmacodynamic effects on the estimation of time by their influence on the activity of the imagination is by no means exhaustive. We have seen (p. 34) that these drugs also have an effect on

Everyone knows about the illusions of time in dreams. A dream which could only have lasted a few seconds or minutes gives us the impression of having continued for a very long time because it contained so many events. Some special cases have shown that there is no relationship between this apparent duration and the actual duration of the images. Let us consider Maury's dream. A book he had been reading left a deep impression on him and he subsequently dreamed that he was condemned to death under the Reign of Terror, spent several months languishing in prison, and was finally guillotined. At this point he woke up with a start and found that one of the rails from the top of his bed had fallen across his neck. The drop of this rail had been the initial sensation which he had interpreted in accordance with the book he had been reading the night before. The dream could not have lasted more than an instant although it spanned a long period of experience (Maury, *Le sommeil et les rêves,* 1861). This is not a unique case. Tobolowska (*Des illusions du temps dans les rêves du sommeil normal,* 1900) gives several similar examples and compares them with the accounts given by people revived after nearly drowning. During the few moments in which they were unconscious they had relived long periods of their lives and, after being saved, they were under the impression that the time during which they were in danger had been far longer than it actually was.

We have all had dreams of apparently long duration when we were only asleep for a short time. This has in fact been tested by experiments. A number of subjects were injected with acetylcholine bromide and lost consciousness for durations of from 4 to 12 seconds. On waking they all stated that they had had very complicated dreams and they estimated the duration of these to be far longer than the time for which they were unconscious (Le Grand, 1949).

It is natural for the mind to ascribe to the dream a duration

---

animals, as regards temporal conditioning. It is therefore probable that pharmacodynamic effects on time judgment are produced in man at several different levels. We should point out that LSD 25 also brings about an overestimation of duration as well as of various other perceptions (Benda and Orsini, 1959).

relative to the events which took place in it (Foucault, *Le rêve*, 1906). But how can we explain the fact that so many images crowd into such a short space of time? There is actually nothing extraordinary in this. Even when we are awake we sometimes imagine in one moment several different consequences of an action we are about to make. Some images are like symbols; they suggest long trains of events or actions which are in themselves quite long, just as a few pictures in a comic strip are enough to suggest a whole adventure. We shall take the example of a dream described by Sturt (*The psychology of time*, 1925, p. 111): "When I was demonstrator in a physiological laboratory I found it hard to wake in the mornings. One day my father came up with a bell and rang it once—up and down—so that it gave two strokes. I dreamed that I was ready to lecture and rang the bell for the body to be brought in. I gave my lecture and dissected an arm, and rang for the body to be removed. The dream occurred between the two strokes of the bell." But it seemed to last about an hour. If we study it closely we will see that it possibly contained only 2 or 3 images. Sturt notes that the lesson actually had no content. The simple juxtaposition of images suggested its contents through a mental construction which took place on waking.

The important fact, however, is the discrepancy which we feel between the apparent duration of the events in a dream—proportional to the richness of the images and their temporal significance—and the actual duration of the dream.

The same facts have been observed in studies on time judgment in hypnosis. If subjects in a hypnotic trance are simply told to wake up after a certain length of time, they do this with far greater accuracy than we are capable of in our conscious time judgments (Ehrenwald, 1931; Loomis, 1951); this reminds us of the accuracy of our physiological clock. If we suggest to the subject, however, that he is going for a walk for half an hour, and then wake him up after 10 seconds, he will describe a long walk and estimate its duration at about half an hour. If it is simply suggested to him that he is performing a task, with no indication of the time it should take, his estimation on waking corresponds more or less to the time which would have been necessary in reality to do this task (Cooper and Erickson, *Time distortion in hypnosis*, 1954). We see in this case again that the time judgment on waking merely corresponds to the number of changes experienced by the subject.

To sum up, when we have to estimate a duration, we can make use of the following information:

1. Quantitative cues which permit a sort of calculation of the duration. When the work carried out can be quantified either precisely or approximately, it can act as a basis for the measurement of duration. This is always very inaccurate as long as we have no instruments available to measure both the duration of the unit of change and the number of changes. These cues permit us, however, to compare with precision the duration of two changes of uniform speed. The estimations thus obtained are objective and we often compare them with those arrived at by use of indications we have actually experienced.

2. Feelings of length of time which can arise during the duration itself through a comparison between the duration we feel and the duration we should like. These have the effect of drawing our attention to every moment of the changes and multiplying the apparent number of these.

3. The number of changes which have been perceived during the activity. This varies greatly in accordance with the attitude of the subject and the nature of the task. The smaller the number of these changes, the shorter the apparent duration. Everything which tends to organize the different elements of the activity into one unit of purpose—structure, significance, motivation—has the effect of reducing the apparent duration.[10]

[10] Piaget distinguishes between two methods of estimating duration: the work accomplished and the activity. Is what he calls activity the equivalent of the number of changes perceived? According to Piaget, activity is the psychological aspect of physical "power," i.e., of strength multiplied by speed (*Le développement de la notion de temps chez l'enfant*, 1946, pp. 50, 285). His definition therefore takes the speed of changes into account rather than their number. In our opinion, however, the speed of changes does not seem to be an essential factor on the plane of perception. We actually do not *perceive* the speed of changes unless they follow each other rapidly; the beats of a metronome are a good example. Most of the changes we perceive succeed each other too slowly for us to have an impression of speed. It is true that we often speak in terms of speed ("How quickly the

These different types of information correspond to different processes and they are not always presented simultaneously. When the work cannot be quantified, no measurement is possible; often the action does not give rise to any feeling of time, but on the other hand the number of changes is always present and imposes on us a very "pregnant" appreciation of duration which resists the—objectively—most well-founded contradictions.

When several of these kinds of information are available, we sometimes systematically disregard certain of them. This may be a trait of character. Some people always seek the most objective signs possible while others rely more on their "feelings." Apart from this spontaneous attitude, it sometimes happens that we are more interested in one aspect or another of a change, according to the situation. If we need precise information, we try to measure the duration as accurately as possible; on the other hand we give more sway to our impressions when we are not under the pressure of time.

These different kinds of information are not, however, mutually exclusive. They may affect and reinforce each other, or again they may clash. We are more aware of the latter case. We are sometimes struck by the contrast when we compare an estimation based on quantifiable cues with the apparent multiplicity of changes or the spontaneous feelings of the length of time. We may also be surprised that we have not been aware of duration when the great number of changes leads us to estimate that it must have been considerable. A day of varied and interesting activities will seem well filled and yet leave us with the feeling that it passed in a flash.

These clashes between information from different sources are shown up by our retrospective estimations of duration. The evolution of memory traces is not the same for the three processes. Although we are very conscious of the quantity of work im-

---

last two hours have gone"), but this is by reference to the periodic movements which measure time. We say just as often: "How short these two hours have been."

We shall continue this discussion later (Chapter 7, p. 240 and Chapter 8, p. 272). At this point we would merely stress that for Piaget the essential seems to be the relationship between the work accomplished and the feeling of activity; for us it is the number of changes perceived.

mediately after the task, it may leave only vague memories later; on the other hand, we may remain far more aware of the number of changes because we felt each of these during our activity. Our feelings of time meet the same fate as all our feelings. We may keep the memory of them—that is, we may know that we had them—but this memory no longer has affective reality, we do not relive it. When I look back some years later on a holiday spent touring, it may seem to have lasted some time, although it was so interesting at the time that it seemed too short.[11] On the other hand my years as a prisoner-of-war seem in retrospect to have no temporal depth, because I have few outstanding memories of that period when every day was filled with gray monotony; yet each day had seemed long while I was waiting for freedom. Of course the actual duration of the touring holiday and the captivity is known, but this knowledge has no direct effect on our intuitive estimation, except perhaps to reinforce it by contrast. This contrast emphasizes the firmness of the basis of intuition for these direct time judgments, founded immediately on the changes we experience and later on the changes we remember.

III *The Relation of Time Judgments to Age and Sex*

Since the appreciation of duration is a result of the integration of complex experiences, it naturally depends on everything which constitutes the personality of the individual. It is apparent from everyday experience that people evaluate the durations they experience in very different ways. The results of all the experiments we have studied show only the central tendency of measurements which actually vary considerably. Unfortunately the study of personality is not yet advanced enough for us to be able to attempt a general differential psychology of temporally organized behavior and in particular of time judgments.

There are, however, a few facts which help toward a comparative study. If we consider nothing but feelings of time, their

[11] Diderot puts this well: "Let us work; work has the advantage, among others, of shortening our days and lengthening our lives." (Quoted by Poulet, *op. cit.*, p. 201.)

presence depends on situations many of which are unavoidable, such as expectation or the continuity of effort, but the frequency of their appearance is not independent of the individual personality. For instance, well-balanced individuals, as we call them, whose motivations are perfectly adapted to the situations in which they are placed, are less likely than other people to suffer from dissatisfaction, because they do not wish so often to escape from their present situation. They therefore have occasion less often to feel that a duration is long. As we have already stated, the very intensity of these feelings of time is a function of the ability to bear frustration or, in other words, of the emotional stability of the individual.

Does the appreciation of duration vary with character or with personality? We have hardly anything to go on for this question. Jaensch and his pupils tried to resolve the problem as part of their work on typology. They found that the *outwardly integrated type*, that is, the individual who tends to interpret what he perceives in a personal way, is most sensitive to the content of a duration when he has to evaluate it; he does not make much attempt to dissociate the objective duration from his impression. On the other hand, the *nonintegrated type*, who analyzes his perceptions without projecting himself on to them, tends to evaluate the duration objectively; he is therefore more accurate. Finally, the *inwardly integrated type*, the introvert who concentrates entirely on himself, has a tendency to underestimate durations (Jaensch and Kretz, 1932; Scheevoigt, 1934). These observations are plausible, but Jaensch's classifications lack objective criteria and it is therefore impossible to use them as a basis for a more thorough inquiry into the problem. We must therefore limit ourselves to a study of how time judgments vary with the more obvious differences attributable to age or sex.

1. The Influence of Age on the Appreciation of Duration

So far we have relied on observations or experiments involving adults, in order to establish the general laws dealt with in this chapter. It would be interesting, however, to follow the development of the appreciation of duration with the growth of the child and to note the modifications brought by old age.

a. The Estimation of Time by Children. When we were discussing the general problem of the appreciation of time, we stressed that the results were always relative to the method (p. 214). This is even more true for developmental psychology where every method of estimation must rely on the "means" which are unequally developed in the child.

It is obvious, for instance, that young children cannot estimate duration in temporal units. Even when they have learned to tell the time, they have no idea of what a minute or an hour represents. They will only learn to use these units with some degree of accuracy after years of experience. At 8 the task is still impossible; at 10 about ⅔ of the children give some reply, but their evaluation of 20 seconds, for example, still ranges from 30 seconds to 5 minutes (Fraisse, 1948). The accuracy of these estimations develops slowly and the training lasts until they are at least 16 (Elkine, 1928).

If this knowledge is not required, however, and if the reproduction method is used alone, children show an appreciation of duration at quite an early age (Fraisse, 1948). In young children, of course, the results are very variable. At 6 years of age, the reproductions of a duration of 20 seconds vary from 1 to 60 seconds; this means that the relative variability is 90 percent instead of 30 percent as for adults. It should, however, be noted that children of that age are very aware of the content of the duration. Filled durations, which in our experiment consisted of a continuous sound, are overestimated, whereas empty durations (the interval between two sounds) are greatly underestimated. The fact of something tangible happening seems to be of prime importance, for the child cannot keep his attention fixed on inward impressions alone.

At 8 years the estimations are far more accurate and less variable. But progress still continues. We asked a group of children to reproduce a duration of 30 seconds; they were to press a button when they thought that a duration equal to a reference had elapsed. We decided to call all those answers *correct* which came within an interval of ±5 seconds. A signal then told the children whether their answer was correct, anticipated, or delayed. Each age group was represented by 20 chil-

dren and each child gave 10 replies; the percentages giving the
different replies were as follows:

Years of Age	Correct Reply, %	Too Long, %	Too Short, %
6	36	43	20
8	45	27	28
10	53	16	30

These figures confirm the observation made above: filled durations are overestimated by young children more than empty
ones. These are particularly indicative of the slow development
of estimation (Fraisse and Orsini, 1958).

According to Gilliland and Humphreys (1943), 11-year-old
children still make errors in estimation almost double those made
by adults. In considering this we should take into account the
fact that these authors combined the three methods of estimation, production, and reproduction. The development of appreciation by the reproduction method alone continues until the
age of 14, according to Jampolsky (1951). It is true that this
author's experiments did not concern simple time judgment. After
tapping several times at regular intervals of 5 seconds, he told his
subjects to continue at the same speed. If we class the subjects in
two groups, those who tapped more quickly and those who
tapped more slowly (the latter group comprising no more than
20 percent of the subjects at any age), we find that the means
for the two groups approach each other as the age increases;
this shows that on the whole the estimations become more
accurate as the child grows older:

	8–9 yrs. secs.	10 yrs. secs.	12 yrs. secs.	14 yrs. secs.	Students secs.
Mean for Those Tapping Too Fast	2.82	3.25	3.05	4.02	4.21
Mean for Those Tapping Too Slowly	8.50	7.61	6.55	5.54	5.75

The child only slowly learns to appreciate duration with the same accuracy as an adult. To what should we ascribe this progress? To the development of new possibilities or simply to a training which takes place gradually through life? Thanks to some recent work by Orsini (results unpublished) we can answer that it is essentially the training which is lacking in a child. Using the same method as that described above (Fraisse and Orsini), she trained 7-year-old children to evaluate durations of 30 seconds, telling them their results each time (correct to within ±5 seconds, too long, too short). This training lasted for 3 weeks. The results show an appreciable and steady progress which was confirmed by tests 3 months later (the children were not expecting this subsequent check).

	Children			Adults
	Before Training	After Training	Three Months Later	Before Training
Correct	9.6%	45.5%	40.9%	36%
Too Long	78.0%	23.6%	29.1%	29%
Too Short	12.4%	30.9%	30.0%	35%

The main problem in genetic psychology is not, however, that of accuracy. It is to know whether children use the same information as adults for their time judgments.

When studying feelings of time, we found that for children the meaning of duration lay in the distance between the awakening of a desire and its gratification. Children have exactly the same feeling as adults, that the time of waiting is too long or that the time of effort, eating for instance, is never-ending. This consciousness of duration appears at the age of about 3 or 4, when the child is becoming capable of accepting the deferment of gratification or of making a continuous effort in order to attain some purpose. Although the actual representation of these feelings of time is less distinct in children than in adults, children are more vividly aware of them and, having less emotional stability, they find it more difficult to bear the conflict between the demands of the present and the satisfaction anticipated for the near future. On the other hand when they are deeply absorbed in the present and are suddenly torn from their spontaneous activity to go and eat or wash, they are more surprised than adults to discover that so much time has passed, and they are even tempted to deny the evidence.

A young child has faith in his feelings of time; this cannot be questioned. But once he has become capable of a certain appreciation of duration, that is when he is 4 or 5, does he also make use of the other two criteria used by adults, the number of changes perceived and the quantity of work accomplished? This question has been discussed most thoroughly by Piaget. Following his research for *Le développement de la notion de temps chez l'enfant* (1946), he came to the conclusion that children estimate duration primarily in accordance with the amount of work accomplished and only at a later stage by the quantity of activity felt. According to him, only older children are capable of dissociating "the work effected from the activity itself and judge the duration according to the introspective characteristics of the latter." (*Ibid.*, p. 50.)

A thorough discussion of Piaget's opinions will follow in the next chapter, for their full importance is seen only in the context of the development of the notion of time in children. Here we shall limit ourselves to a few factual observations. According to our own experience, young children of about 5—that is, the youngest children who can be given simple problems to check their methods of time judgment—estimate durations sometimes by the work accomplished and sometimes by the changes perceived. To prove this we shall use our own experiments, but also some of Piaget's, for some of his work can be interpreted in accordance with our ideas. Although the distinction between the two successive stages appears to be clear to Piaget when the child is estimating what he calls "physical time," the duration of changes taking place around us, he also admits that in the estimation of the duration of action "there is far greater continuity between the reactions of young children and those of older children, and the illusions which play a part in the evaluation of these durations take a qualitative form which is common to children and adults." (*Ibid.*, p. 242.)

All the experiments we are about to describe are organized on the same principle. The child is set the task of comparing two periods of time during which the changes which take place and the work accomplished are of a different nature. From his reply and in the second place from the reasons he gives, we shall see on what information he bases his estimation.

We shall first consider the case of the appreciation of the duration of an action based on an experiment by Piaget (*ibid.*, pp. 253-256). The child's task was to transfer small pieces of wood in one case and lead in the other, from one box to another, using a small pair of tongs. Unknown to the child, the duration of the two tasks was the same. It was obviously easier to transfer the pieces of wood, and the child therefore transferred more of them than of the pieces of lead. When they were asked questions afterwards, a few of the children thought that they had spent longer transferring the pieces of wood, "because I moved more," but these were a minority. Their judgment was founded on the work accomplished: more pieces moved = more time. Other children found that, on the contrary, more time had been spent transferring the pieces of lead. Their reasons were clumsy: "Because it is bigger," "It is heavier," "I got stuck." But they emphasize the difficulty of the task. As we have already seen, this difficulty has the effect of drawing the attention of the subject to every change. Each piece transferred "counts." This is also true of a difficult walk in the snow, for instance. Piaget remarks very aptly that "when you are walking in mountainous country and the snow comes to above your knees, ten minutes of effort and slow climbing seem like at least twenty, whereas an easy pace leads to normal judgments." (*Ibid.*, p. 259.) Some of the children even said that they had transferred more lead than wood; this is objectively false but it corresponds perfectly to the number of changes noticed. This analysis was confirmed by older children who said that they had to move the pieces of lead "one by one." We can therefore assume that the children who judged the time spent transferring lead to be longer based their estimation on the number of changes.[12]

[12] This is our interpretation. Piaget thinks that children always judge according to the work accomplished, not only in the first case, on which we agree, but also in the second, because the lead gives the impression of "more work." In our opinion the word *work* has a different meaning here; it relates to the activity, that is the number of changes felt (see Chapter 8, p. 275).

The same conclusion is reached as the result of a very different experiment which we carried out using another method borrowed from Piaget (Fraisse and Vautrey, 1952b). This time the child is set the task of comparing the duration of rectilinear movement of two lead toys (racing cyclists) which move on parallel lines on a table, in the same direction and at the same time. An adult can easily decide on the relative duration of movement if he knows the order of departure and arrival, the speed and the length of the course, but we shall see in the next chapter that a child is not capable of such calculations until the age of about 7 or 8. In this case, how does a child of 4–5, with no means of measuring the duration, solve a problem which requires a direct appreciation of time? Before describing this experiment we should point out that our aim was not to find out if he could give the correct answer but to discover what means he uses to make his estimation.

Two cyclists A and B leave the same line simultaneously and move in the same direction. A moves twice as quickly as B. The two figures stop simultaneously; they have moved for the same length of time but A has covered twice as much ground as B. Only 17 percent of the children find that the durations were equal. The others fall into two more or less equal groups. Some think that A moved longer and when asked why, they refer to the greater speed or the greater distance covered. Their error in judgment is due to the fact that they base it on the work accomplished: A achieves more than B. The other children think that B was moving longer. Their explanations show us why. "He was tired," "he couldn't be bothered;" it therefore seems that in this case the children judge the duration of movement of the cyclist by identification with their personal experience. When you fall behind the others and go slower, it is because the task is too difficult, and when this happens you take more notice of every movement which costs an effort. The child who judges that the slower cyclist, who did not cover so much ground, was moving for a longer time than the other, bases his estimation on the same criterion as the child who thought he had spent longer transferring pieces of lead than pieces of wood. They both judge the duration according to the changes experienced, one directly and the other by identification.

This interpretation of the children's replies is confirmed by other experimental situations in which the durations of movement are unequal. In these circumstances the majority of children manage to give the correct answer, implying—in the absence of operational comparisons—that they make accurate use of information corresponding to reality. Let us consider in turn the two cases which may arise. In the first case, the figure which moved for the longer time was the one which "did more;" for instance, the two cyclists leave the same point simultaneously, moving in the same direction at the same speed, but one moves for twice as long and therefore covers twice as much ground as the other. Seventy-one percent of the children recognize the fact that he moved for a longer time. Much the same results ~re obtained if the speed of the cyclist moving for the longer time is also greater.

In the second case the cyclist who moves for the longer time does less than the other; he covers less ground or moves more slowly. If the children were capable of judging only according to the work accomplished, they would all make the same mistake, but this does not happen. Let us suppose that the two figures leave *one after the other* from the same point but the second figure moves faster than the first and they arrive together at their goal. This is the story of the hare and the tortoise: 61 percent of the children think that the *slower* figure moved for the longer time.

The reasons given by the children confirm that in the first case they were guided by the quantity of work accomplished by the moving figures, but in the second case they were guided mainly by the quantity of change during the action. The "tortoise," which went slower, had more difficulties. The majority of children are therefore capable of choosing the *right* criterion, but a considerable minority are grossly misled by their trust in another one.

It is apparent, then, that young children use the same information as adults but their estimations have certain peculiarities which can be seen from the experiments described above. The first of these is that children use direct, comprehensive means of estimation more than adults do. The adult knows that his judgments are not absolutely reliable because he has often experienced this. Therefore he tries whenever he can to evaluate durations indirectly. In the case of a race, for instance, he calculates according to the points of departure and arrival or the relationship of space to speed: the faster moving object takes less time. When watching a race, whoever would think of making a direct judgment on the time taken by the runners when he *knows* that the winner took less time than the runner-up? We are content to note the order of arrival and we infer the duration through implicit reasoning. But as we shall see in Chapter 8, children are not capable of such deductions; they rely entirely on more direct methods of appreciation.

The second peculiarity is precisely this faith which the child has in the method of appreciation he uses. He centers himself on it. He does not weigh his judgment against an attempt at measuring. He is also not aware of the number of possible methods of estimation which produce for adults the phenomena of contrast that we stressed earlier. Piaget's experiments are all very striking from this point of view. What he calls the stage of articulate intuition, intermediate between the intuitive and operational stages, could be fairly well described by saying that the child begins at this point to doubt his first intuition and compare it with the other means of estimation at his disposal. Piaget notes this increasing caution of children who, as they grow older, use expressions such as "it seems to me" more and more when the situation does not permit of exact measurement; this is particularly true when they compare the durations of transferring the pieces of wood and pieces of lead.

What determines the child's choice of one type of information rather than another? When we were speaking of adults, we noted that typological differences must explain the fact that some people are more aware of the work accomplished and others of the changes experienced. Is the same thing true for a child? No attempt has been made to check this. We have only observed in our experiments that the same children based their appreciations indiscriminately either on the work accomplished or on the effort felt. It is true that they were presented with ambiguous situations; we intended them to be so in order to show the possible types of answer. In ordinary life it is of course the nature of the situation itself which determines the kind of information used. When we are estimating the duration of a physical change, we mostly have at our disposal information relating only to the quantity of work, but when we are judging our own actions, the changes we feel are frequently the basis of our estimation.

The third peculiarity is a result of the way in which the child quantifies the work accomplished. From this point of view he is also content with general intuition and not with an estimation which takes all the facts of the situation into account. Let us take an example from Piaget's work (*op. cit.*, p. 130), since he stressed this aspect. Water flows from a flask down a Y-shaped tube, regulated by a single cock, into two vessels of different shape and volume. If the water is allowed to run until the bottle with the smaller capacity is full, the second of course still being only part full, a young child will judge that the water flowed for a longer time into the full bottle and that this contains more water. The child's error concerning the "work accomplished" influences his appreciation of the duration. A slightly older child will not be taken in by this perceptual intuition. He will be capable of considering all the facts of the situation, the synchronization of flow and the equality of volume (see Chapter 8, p. 266). Piaget also showed that children often make mistakes by confusing greater speed with greater duration. The object which moves faster does more work, and because he has not learned to relate speed to space the child mistakes the amount of work accomplished.

Because they are incapable of using actual measurements in their estimations of duration, children are even more dependent than adults on what takes place in this duration, whether it is physical changes which he observes with his own means of estimation, or the changes he feels.

Even when a child has reached the age of 7 or 8 and is beginning to be capable of reasoning and to have some notion of time, he still seems to be more sensitive than adults to the qualitative aspects of the duration. If his judgments become more accurate as he grows older, it is probably because he makes better use of the units of time but also because he has less and less faith in his immediate impressions. This development is shown well in the work of Axel (1924) to which we have already referred (p. 225). He asked a number of children aged from 9 to 14 to estimate in minutes and seconds the duration of various tasks: empty duration, writing letter Is, crossing out signs, and doing mental arithmetic (mathematical progression in 7s). The following table gives the means of these time judgments and shows the salient trends (duration of every task: 20 seconds):

Age	Number of Children	Empty Duration	Writing Is	Crossing Out	Mental Arithmetic
9	60	120	70	45	27.5
10	135	46.5	35.5	20	20
11	117	32	23	15	17
12	93	30	22	15	15
13	122	31	22.5	20	15
14	138	30	18.5	17	13

Our first observation is that the durations are very much over-estimated at the age of 9 but become more and more accurate as the age increases, until they approach the actual duration. The main point to notice is that at every age the tasks fall into the same order, as regards their apparent duration, as for adults. However, the discrepancy between the estimation of the tasks which seem longest and that of the apparently shortest decreases with age whether this is considered as an absolute or, more accurately, as a relative difference. Thus at 9 years of age the task of writing Is seems 2.5 times as long as that of mental arithmetic but at 14 it is only 1.4 times the length; there is a gradual and general evolution.

To sum up, children base their time judgments on the same information as adults but they have not yet learned to relate the various possible estimations of duration. As they grow older they will learn to compare these with, and correct them by, indirect evaluations based on methods of measurement: the order of succession, temporal cues, the inverse relation of time to speed.

b. The Estimation of Time in Old Age. It is obvious that the problems connected with old age are of an entirely different kind. From the evolution of mental functions we may not conclude that they return to nothing, and there is nothing to show that old people do not appreciate duration in the same way as adults. But there is one point which has often been made: "The same space of time seems shorter as we grow older." (James, *Psychology, briefer course*, 1892, p. 284.) But James himself restricts the application of this law. It is valid for the appreciation of days, months, and years, but certainly less valid for hours. This has not been checked by any experiments; it would be difficult to do so, for it concerns the absolute impressions of old people. If they were asked actually to estimate a duration, they would probably be able to correct their initial impression and give just as accurate an answer as younger people.[13] The fact noted by James has, however, been observed so often that it is probably true. We are even tempted to think that it is just as applicable to hours, that is, to time which has just been lived, as to longer durations. The phenomenon is merely less apparent because it concerns a smaller length of time.

This fact is most frequently interpreted in a way which concurs with our idea that the estimation of duration is relative to the number of changes experienced. James actually explains the brevity of time by the fact that in old age one is so used to the events of life that they do not leave individual memories. Guyau says the same thing: "The impressions of youth are vivid, novel and numerous; the years are therefore filled and marked in a thousand ways . . . the backcloth of the stage retreats behind the scenery which changes constantly in full view of the audience. . . . On the other hand, old age is the scenery of a classical play, always the same, an unremarkable décor. . . . The weeks are alike, the months are alike; this is the monotonous pace of life. All these images are superimposed on each other and merge into one." (Guyau, *Genèse de l'idée de temps*, 2nd ed., 1902, pp. 100-101.)

[13] Although not really dealing with old people, Pumpian-Mindlin (1935) found that people aged from 40 to 60 did not make any greater systematic errors in their estimations than subjects aged from 20 to 30.

Old people are less aware of change because they live a tranquil existence, but even more because they are less aware of changes in the course of their activities as they have grown too accustomed to them. This is exactly the same phenomenon as that which we have already analyzed (p. 219): when we repeat the same task again and again, the time it takes seems shorter and shorter because our mind no longer needs to pay so much attention to every moment of the action. This absolute brevity of time may explain why time seems to pass relatively more quickly: it introduces a contrast between experienced time and time as measured by clocks and calendars.

Other interpretations have been suggested to account for this apparent acceleration of time with age; they do not contradict the ideas we have already expounded. Paul Janet (1877) suggested that the apparent duration of a period of time may be relative to the total duration of one's life. At 20, 1 year represents $\frac{1}{20}$ of one's life but at 60 it is only $\frac{1}{60}$. It is probable that this relationship does play a part in our evaluation of the periods of our life, for our judgments of a part always bear a relationship to the whole to which it belongs and Weber's law, in its wide sense, may also apply in this respect (Benford, 1944). It is obvious that one year counts more to a child than to an adult or an old person. Our birthdays become less important as we grow older; children never forget how old they are but an adult often has to make an effort to remember or calculate. But this law can only be applied to long durations when perspective, at least implicitly, has some part to play. It cannot be said that old people judge the length of each day according to the number of days they have lived.

More recently, Carrel (1931) and Lecomte du Nouy (*Le temps et la vie*, 1936) suggested another interpretation. Having discovered that wounds heal more slowly as we grow older, they deduced from this a physiological time which cannot be assimilated to astronomical time. The number of biological changes which take place in one unit of sidereal time is greater when we are young than when we are older, which means that more work is accomplished by the organism in the former case. "At different ages we need different quantities of time to accomplish

the same work, the healing of a square centimeter of wound."
(Lecomte du Nouy, *ibid.,* p. 234.) To reverse the situation, if we
take the duration of a biological phenomenon as the unit of time,
for instance the duration of the healing process for one square
centimeter of wound, sidereal changes of increasing size will
correspond to this unit as we grow older and we will say that
time is passing more quickly. Lecomte du Nouy assumes that our
psychological appreciation of time is itself related to the biological unit of time, the "subconscious providing our intelligence
with unprocessed information." (*Ibid.,* p. 237.)

But is there a direct relationship between biological and
psychological time? We are justified in thinking that biological
time has some influence on our estimations of duration. In fact
we saw earlier that in animals and humans these estimations are
dependent on the temperature which activates or slows down
biological exchanges.[14] But what connection do these facts have
with the appreciation of time? It is quite possible that in old age,
decreased biological activity makes us register fewer changes
and, by the same token, days and hours seem to be shorter by
contrast than before. In our opinion, however, these are only
basic tendencies which are corrected, all through adult life, by
our social habits. Although old people all agree that time passes
more quickly than it used to, it is none the less true that there is
hardly any change in their objective time judgments. Psychological time is doubtless conditioned by biological time, but
one cannot be equated with the other, for the psychological
processes are more complex,[15] since they involve all the functions: the phenomena of contrast, automatic corrections, habit,
etc.

[14] These exchanges are not to be confused with the simple acceleration
of physiological rhythms such as those of the heart and respiration. Schaefer
and Gilliland (1938) have shown that such acceleration does not modify
our estimations of time.
[15] This is probably why Gardner (1935), who also started out from a
physiological hypothesis, did not find any difference between time judgments in cases of hyperthyroidism and hypothyroidism. The experience of
time of these subjects may be different but corrections of social origin influence their estimations.

2. The Influence of Sex

The influence of age on time judgments is apparent to everyone, but it is mainly through experimental situations that it has been found that there may be a difference between men and women in this respect.

The problem was raised by MacDougall in 1904. He found that four different tasks were, on an average, overestimated to a greater degree by women than by men. Yerkes and Urban studied the same question in 1906 and obtained the same result. Axel (1924), in the important research work that we have already quoted several times, also found that women estimated durations to be longer than they really were, whatever the task involved. We give below the means for the estimations of 46 men and 42 women, for tasks of 30 seconds' duration.

	Men	Women
Estimation of an Empty Duration	27	36.5
Tapping	26	38
Crossing Out Signs	20	32
Finding Analogies	18	25.5
Completing Series of Figures	14	24

Gulliksen (1925) used larger groups of subjects and also found that the durations of *all* the tasks were estimated to be greater by the women than by the men; these differences were significant except for the most difficult tasks (reading in a mirror, dictation, divisions).

This law appeared to be firmly established, although no explanation had been suggested, but it has been queried by more recent work which cannot be considered any less thorough than the above research. Swift and MacGeoch (1925), for instance, did not find any difference between the estimations of various tasks by men and women, in fact the women showed if anything a tendency to overestimate the durations relatively less than the men. Harton (1939) made his subjects estimate the duration of periods of 4 minutes devoted to four different activities and found that the average estimation of the men was 287 seconds ($\sigma = \pm\ 101$) and of the women 243 seconds ($\sigma = \pm\ 76$). Gilliland and Humphreys (1943) combined the results obtained by the three methods of estimation, production, and reproduction and found no difference between the sexes. A critical study of the methods used does not explain these divergences. We shall therefore not attempt to explain the possible differences between the time judgments of men and women. More experiments will have to be made in varied conditions to settle the question before it can be interpreted. We discussed it here only because we thought this preferable to passing over it in silence.

III

THE NOTION OF TIME

"The notion of time, like that of space, is empirically the result of the adaptation of our activity and our desires to an unknown and perhaps unknowable environment."

Guyau, *Genèse de l'idée de temps,*
2nd ed., 1902, p. 46

THE TYPES OF TEMPORALLY ORGANIZED BEHAVIOR WE HAVE SO FAR discussed have all concerned isolated sequences of change.

No conditioning to time can take place unless the series of changes concerned affects the organism: the synchronization of the rhythms of organic activity to the circadian rhythm, adaptation to the duration which separates a conditioned stimulus from the stimulus it announces.

The perception of time is only possible through the organization of stimuli, and for this the changes which form the stimuli must have a certain homogeneity; this means in practice that they must belong to the same sequence.

By forming representations of changes, a man can widen the field of his behavior. He can acquire a temporal horizon and have a basis for certain kinds of time judgment. But if considered in isolation, these types of behavior are confined within narrow limits. To see this we only need to consider the activity of a child, which is subject to a double restriction; he can only grasp the order of simple series of events by intuition and he only apprehends duration through the changes he has felt.

These limitations will be our starting point for a study of the progressive development to the most perfect form of adaptation to change, the notion of time. By considering the genetic stages in this progress we shall define its nature and the part it plays in our conquest of the universe.

I Limitations in Temporally Organized Behavior at the Prenotional Stage

The best way to show the possibilities opened to man by the notion of time is to consider what he does, or rather what he cannot do, at an age when he does not yet have this notion. At the age of 5,[1] a child already has temporal perspectives; he has become conscious of the resistance of time and he is capable of simple estimations of duration. He still fails, however, in many of the temporal problems posed by life.

The changes amid which we live are defined by the order of their succession and the intervals of duration between them; if we can grasp this order and the intervals between changes, we have all the elements necessary for the reconstitution of the series. How does a child arrive at this dual knowledge?

We shall consider in turn the problems raised by the order and by the intervals.

1. The Apprehension of Order

A 5-year-old child has no difficulty in perceiving the temporal succession of two events, but this perception is very precarious if the two stimuli do not belong to the same series of events, for he is not capable of confirming this fleeting information by the reasoned used of other cues.

For the following example, and for many others in this chapter, we are indebted to the work of Piaget, *Le développement de la*

---

[1] We use a 5-year-old child as a reference in this section because at this age his verbal and general development already permits of serious experiments and yet he is still not capable of reversible operations; these are not possible until the age of about seven.

notion de temps chez l'enfant (1946). His analyses and experiments handle the problem with such originality and penetration that any subsequent discussion of the subject cannot but refer to his work.

A child is shown a race between two toy figures, one yellow and one blue, which are moved along parallel lines in the same direction on a table. They start simultaneously but move at different speeds. The yellow figure moves faster, goes farther and stops first; the blue figure moves more slowly and goes on moving a little after the yellow one has stopped; he therefore stops after the yellow figure but has not caught him up in space. The following dialogue takes place with a 6-year-old child: "Did they stop at the same time? *No, the yellow one stopped before the other.* Which one stopped first? *The blue one.* Which one stopped sooner? *The blue one.* What do we do at midday? *We eat.* Now if we say that the yellow one stops at midday, when does the blue one stop? (The race is shown again.) Does he stop at midday too, or before midday or after midday? *Before midday.* Look (the race starts again). *Yes, the yellow one stops first. He moved for longer.* And the other one? *He stops before midday. . . .*" (*Ibid.*, p. 91.)

The child perceived correctly that the yellow figure stopped first, but the differences in speed, distance covered and final position interfere with this fleeting perception and confuse him. He is particularly confused by their order in space and he interprets the spatial delay as a temporal delay. The blue figure did not move so far, and so he stopped before midday. This mistake is not made if the temporal order is confirmed by the spatial order. If the child were asked about the successive positions of one and the same figure he would not make a mistake; even if there are two movements in opposite directions, no mistake is made. In this case the two movements are dissociated and the perception of the temporal order is not confused with anything else and is therefore not doubted; Piaget therefore concludes that the confusion in the case of two movements in the same direction is not verbal, since the term "first" can have a temporal or spatial meaning, but is the result of a logical confusion between the facts observed in the experiment (*ibid.*, p. 92).

The same phenomenon is to be found in connection with the perception of the nonsuccession, or simultaneity, of two events. If the same two figures, yellow and blue, stop simultaneously, but the yellow figure has moved farther than the blue, the child recognizes the fact that the yellow figure does not go on moving when the blue one stops but he does not agree that they stopped at the same time, because the yellow figure has moved farther than the other (*ibid.,* p. 106). This mistake is not made if the yellow figure and the blue one stop on the same line; there is then no difficulty in the way of the child's seeing that they stopped simultaneously. Once again, the temporal order is not dissociated from the spatial order, but if the two coincide, no mistake is made.

In what way is the adult superior? If he doubts his feelings, he can use reasoning. To use the same example, an adult can *reconstruct* the temporal order by considering the positions of departure and arrival and the speed at which the objects were moving. This reconstruction is even more necessary when the order of events is not the object of a perception but of a recollection by memory. The memories of children are completely jumbled up, for they have not learned to reconstruct their past; this is shown by the haphazard way in which they retell stories, for the order of events depends more on their interests or on incidental associations than on reality.

Using another technique, Piaget studied the difficulties peculiar to children in the determination of succession; he made them reconstitute a story from a number of pictures arranged in random order. This method shows exactly where children go wrong. The order established by the child is in fact fortuitous; he tries to justify it by a syncretism which corresponds to no particular order, whether chronological, causal, or deductive. Because this order is intuitive, children aged 5 or 6 cannot modify it, even if they are made to see that they are wrong. They would have to introduce causal relationships, founded on probability, and this would require either knowledge or experience. They would in particular have to trace the cause from the effect, which implies the ability to run through the series in both directions; they would, in fact, have to be capable of reversibility. But perceptual intuition is characterized by irreversibility.

This same limitation is illustrated by Wallon's dialogues with young children. "Has the Seine always existed? *Yes.* Has Boulogne always existed? *Yes.* Which came first, the Seine or Boulogne? . . . Which existed first? *Boulogne.*" (*Les origines de la pensée chez l'enfant,* vol. II, p. 211.)

This kind of mistake is definitely a result of the child's lack of knowledge. The nature of the mistake, which often occurs, brings Wallon to the conclusion that the child considers the town to have existed before the natural phenomenon because "houses seem to be a necessary part of the child's existence whereas the river or lake belongs more to accompanying circumstances." (*Ibid.,* p. 212.) In the same way a child will sometimes bestow on himself an absolute or relative antecedence, even in relation to his own parents.

In the case of events which they have not lived themselves, it is experience which the children lack. When a child has this experience, he intuitively knows the order of events without checking it by causal relationships or logical sequences. But his apprehension of duration is still very inadequate.

It is fairly obvious that children are generally incapable of reasoning based on the facts they have experienced. In the case of time, this inability is apparent specifically when they are faced not with a simple seriation—permitting the beginning of a temporal horizon—but with the coseriation of several series of events. The difficulties they encounter will be seen more clearly when viewed, with Piaget, from the level of our adult experience: "When we think of series of events in our own past which are independent yet interrelated (e.g., dates connected with the administrative aspect of our career, order of publications, private life and succession of political events), we find that although these series all remain very vivid in our memory, we cannot, without recourse to logical and operational reconstitution, (1) say whether a particular event from one series occurred before or after another from a different series (although we are well aware of the order of succession in each series), or (2) evaluate approximately (by + or −) the respective durations which have elapsed between events belonging to different series. . . ." (Piaget, *op. cit.*, pp. 265-266.)

Obviously we cannot group events from two different series in relation to each other unless we know not only the order for each of the series but also the exact duration between events in both series. A child, limited to general intuitions, is incapable of this exact estimation of durations.

Furthermore, at the preoperational stage of thought, a child is not capable of interrelating successions and durations in order to reach an over-all reconstruction of the order of several series of changes. We shall consider these two points in turn.

2. The Apprehension of Duration

We shall begin by recalling some of the conclusions reached in Chapter 7. At 5 years of age a child has intuitions of duration, in fact he is even capable of estimating duration by using the same criteria as an adult. But there is one most important difference: his immediate impressions are not corrected by other information which would enable him to evaluate time in units of measurement. The result is that the estimations of a child are very similar to those of an adult in situations which provide only introspective criteria, but they appear more and more incorrect as the situation permits of a reasoned evaluation of duration.

Let us take a simple example: for the child as for the adult, 15 seconds spent standing with folded arms seems longer than 15 seconds spent looking at an amusing picture (Piaget, *ibid.*, p. 257). The only means of comparing these two durations experienced one after the other are feelings of time and the number of changes felt. The adult is in exactly the same position as the child; both have the same "illusions."

Let us take a more complex example. A child is given the task of drawing lines for 15 seconds, first carefully and then a second time as fast as possible; young children agree unanimously that the time was longer when they were drawing more quickly (Piaget, *ibid.*, pp. 241-250). This is not surprising: the information based on the work accomplished (more lines drawn in the second case) concurs with that provided by the changes felt (when an effort has to be made, every moment counts). These evaluations founded directly on activity do not vanish as the child grows older. According to Piaget, a third of the children between the ages of 10 and 13 still make the same mistake as very young children. But the dawn of reasoning gradually breaks in older children; when we go faster we do more things in the same time. When they realize this, many children mistrust their intuitive estimations and infer that the two tasks may have had the same duration.[2]

When it comes to comparing more or less simultaneous durations, however, the estimations of a child are quite different from those of an adult. When describing experiments of this kind in Chapter 7 (the two cyclists, pp. 241 et seq.), we stressed the fact that children use the same information usually employed by an adult; but the child is confused by this information whereas the adult seeing this experiment is not misled. Both have the same impressions, which can give rise to the same illusions, but the adult does not stop here, he uses his reason and tries to measure. It is these operations which a 5-year-old child is incapable of, because he cannot form the representation of a duration independent of its contents. In other words, he is incapable of forming a representation of a homogeneous duration and of temporal intervals independently of what happens in them. If different events take place in two equal durations, whether simultaneous or successive, the child cannot grasp the identical temporal nature underlying the appearances.

In short, he is incapable of measuring duration. Measurement requires the knowledge of a unit. In time, this unit can only be a uniform change which serves as a reference, but this uniformity is a construction, not a reality. In order to use the sidereal day as a measurement, men postulated that the changes were uniform because they had observed that similar effects recurred between the periodic phenomena. Even when the uniformity of a change is *perceptible*, like the trickling of sand through a sandglass or the turning of the hand round a stop watch, it is not a stable fact. As Piaget points out, even adults make this mistake. "When we

[2] The same evolution with age is seen in the experiment with pieces of wood and lead (see the description of this experiment, Chapter 7, p. 240).

watch the sand run through a sandglass beside the telephone while making a long-distance call, or time an interesting race or the delayed reaction of a subject in an experiment, we may also have the perceptual illusion of a change in the speed of the sand or the hand of the stop watch, and have, as the case may be, a positive illusion or an illusion of contrast." (*Ibid.*, p. 188.) This is even more true of a child. The sand seems to flow faster when he works faster and slowly when he works more slowly; sometimes the reverse happens because the effect of contrast is predominant (Piaget, *ibid.*, p. 186). But in the child, the first impression is most important; he really believes that the sand or the hand moves faster or more slowly whereas the adult knows it is an illusion. "We know that the movements are constant and we therefore attach no importance to the perceptual aspect of such observations; at the most we are amused by the apparent resistance or cold irony of these mechanisms which oppose our will." (Piaget, *ibid.*, p. 188.)

Actual measurement will not be possible until the child recognizes the uniformity of the change independent of his impressions and when he acknowledges the existence of homogeneous time independent of the content of the action or events which fill it; in other words, when he is capable of forming an abstract representation of duration.

3. The Independence of Order and Duration

The young child's grasp of the *order* of events is incorrect, or at least uncertain, as soon as the perception of succession is not obvious and, even more, when he has to reconstitute it later. The *durations* between the events are only appreciated in accordance with their content; this means that they give rise to many illusions.

The difficulties encountered by a child on these two planes lead him to make many mistakes. He would often be able to overcome these if he were only able to relate the information regarding duration and order, consider one or the other according to the circumstances, and check or supplement his over-intuitive information, since order and duration are logically complementary.

The inability of the child to relate order and duration is particularly apparent when he has to determine the relative age of two people. Piaget's work is again most helpful. The problem is difficult in itself. When determining the relative age of two people, we have neither experience of the order of their births nor the possibility of estimating the duration of their respective lives. The adult can rely on his knowledge; he either knows their dates of birth or if not he can interpret the signs of growth in a child or of aging in an older person (wrinkles, gray hair, heavy gait); in other words, he can judge what point in life each person has reached. In both cases a deduction is necessary to fix the relative age of the two people. Inversely, if the adult knows their ages, he can deduce the order of their births. Children are incapable of such operations. They usually know whether their brothers, sisters, and friends are older or younger than themselves, but they cannot deduce from this the order of their births. Rom (aged 4 years, 6 months) has a younger sister called Erica: "Who was born first, you or Erica? *Don't know*. Can we find out? *No*. Who is younger? *Erica*. Well then, who was born first? *Don't know*." (Piaget, *ibid.*, p. 211.) When they do not know someone's age, children rely on the signs of growth and deduce age from size; this deduction is often correct in the case of children, but it still does not help them to make any conclusions regarding the order of birth. This leads to mistakes, the most obvious being that all adults are of the same age because they are the same size. Let us quote another conversation which illustrates all these inabilities. Andy (aged 6) has a friend, "Younger or older than you? *Older*. Was he born before or after you? *After*. Is your daddy older or younger than you? *Older*. Was he born before or after you? *Don't know*. Who came first, he or you? *Me*. Do you always stay the same age or do you grow older? *I get older*. How about your father? *He always stays the same age. . . .*"

Rather than continue with these conversations which are sometimes difficult to interpret, let us quote an experiment described by Piaget which shows that children are incapable of drawing conclusions as to relative age even if they know the order of birth. The method used consists in giving the child two series of drawings representing orange trees and plum trees. It is explained to him that in each series the pictures are of the same tree, photographed every year. At one year of age there was one fruit, at two there were two, and so forth. The child is then given the pictures of the orange tree and he has no trouble in putting them in the correct order. He is then told that when the orange tree was two years old (Or_2) and had two fruits, a plum tree was planted; then the plum tree with one fruit (Pl_1) is placed under Or_2, Pl_2 is placed under Or_3, Pl_3 under Or_4, etc. The youngest children cannot even conclude with regularity that Or_2 is older than Pl_1; at 6, 50 percent only conclude for each *pair* of pictures that the tree with more fruit is the older.

Even when it is possible, this conclusion is the result of a simple quantitative intuition and has no reference to the order of planting of the two trees. Piaget proved this by another experiment in which the speed of development of the two trees was not the same. The quantitative signs therefore did not bear a direct relationship to the age of the trees and the children's replies depended on the development of the tree and not on its actual age. In this experiment Piaget used apple trees (A) and pear trees (P). Each drawing consisted of larger and larger branches holding larger and larger circles which contained more and more apples or pears. But the pictures also showed that the pear tree was developing faster; the drawings of the apple trees ranged from A_1 (13 millimeters diameter, 4 apples) to A_6 (80 millimeters diameter, 44 apples), while the pear trees went from P_1 (12 millimeters, 4 pears) to P_5 (99 millimeters, 74 pears), so that A_4 (60 millimeters, 27 apples) equalled P_3 (60 millimeters, 27 pears).

As in the preceding experiment, the pictures of the apple trees were first placed in order, and it was then explained to the child that when the apple tree was two years old a pear tree was planted, which was then one year old, and that the two trees were photographed every year. The pear trees were therefore placed under the apple trees: P_1 corresponding to A_2, P_2 to A_3, etc. The young children were able to see which was the older tree as long as the order of planting corresponded to the development, but they failed as soon as the pear tree had outstripped the apple tree in size. The following conversation is very revealing: "Jock (aged 5 years, 6 months) succeeds in arranging the apple tree pictures in order, saying to himself *1, 2, 3,* etc. Look, when the apple tree is 2 years old a pear tree is planted. Which tree is older? *The apple tree.* And the next year? *Still the apple tree.* And here are the photos taken the next year on the same day of both these trees ($A_4 = P_3$). Which tree is older? *The pear tree.* Why? *Because it has more pears* (this is false; both have 27 fruits). And here (A_5 and P_4)? *The pear tree.* How old is it? (Jock counts one by one.) *Four.* And the apple tree? (He counts pointing to the pictures in turn.) *Five.* And which of the two is older? *The pear tree.* Why? *Because it is 4.* Are you older at 4 or at 5? *At 5.* Then which is older? *I don't know. . . . The pear tree, because it has more fruit.*" (Piaget, *ibid.*, p. 229.)

As long as this child could check the order of birth by the development in size and quantity of fruit, he made no mistake, but as soon as the pear tree grew larger than the apple tree and had more fruit, he was misled because he continued to use the same criterion, not knowing how to check this against the order of birth, despite the hints from the experimenter.

Two other conversations with older children show even more vividly this limitation of the child who cannot yet reason. In the case of the first child, Pip, the intuition of age based on development is coexistent with the knowledge of the time of planting; in the second, Paul, the integration is complete: Pip (aged 6 years, 8 months): "P_4 is older than A_5? *Oh no, the apple tree is older because it is 5 and the pear tree is only 4.* And this year (P_5 and A_6)? *The pear tree is older because it has more fruit.* Are you sure? *Oh no, it's the apple tree because it is 6 and the pear tree is 5.*" (*Ibid.*, p. 231.)

Paul (7 years, 2 months): "(A_5 and P_4.) *The apple tree is older because it was planted first.* And A_6 and P_5? *The same. It makes no difference that it is larger; I have a friend who is bigger than me and he is only 6.*" (*Ibid.*, p. 232.)

These examples all show what difficulties hinder the 5-year-old. He cannot go beyond intuition and immediate appreciations. We must now try to show the stages through which he becomes capable of adapting to *all* kinds of temporal information.

II *The Development of the Notion of Time*

As he grows older, the child's progress is marked by two successive stages. In the first stage there is an evolution in his intuitions of order and duration, which become more and more independent of his immediate, concrete experience thanks to the development of the corresponding representations.

But these representations do not yet permit him to relate information concerning order and duration. The child develops this ability at a later stage through reversible constructions, or, in other words, through operations.

1. The Evolution Toward Representations of Order and Duration

a. Order. As we have seen, order presents no difficulties for a young child when it is perceived in unambiguous conditions. But if the temporal order must be distinguished from the spatial order, confusion ensues. As soon as the order has to be recalled rather than perceived, young children fail because they are not yet capable of reconstituting the sequence of their memories.

Piaget has shown that this inability to reconstitute the order of images or memories is bound up with the syncretic nature of the perception or of the mental image and that the child makes the essential progress when he becomes capable of detaching himself from his immediate intuition and makes *assumptions* concerning the actual order of succession. This implies that the child is capable of forming a *representation* of a series of events and also that he can give a meaning to this succession, or in other words reconstitute the logical seriation of events, particularly with regard to causality.

This stage is perfectly illustrated by a very apt experiment of Piaget's. The child is shown two bottles of different shape placed one on top of the other. At the beginning of the experiment the lower bottle is empty and the upper one is filled with a colored liquid; a certain quantity of liquid, always the same, is allowed to drop from the upper bottle to the lower one at regular intervals, by means of a glass cock. In front of the child there is a series of drawings representing the two bottles. Every time the liquid flows down, he is asked to mark on a fresh drawing the level reached by the liquid in each of them. When all the liquid has dropped into the lower bottle, in 6 to 8 stages, the child is asked to arrange the drawings in order, starting on the left with the drawing he marked first "when the water was all at the top" and so forth.

Young children (of about 5) fail in this arrangement of the drawings. They are, however, capable of showing the successive levels of the liquids on the bottles themselves, for they rely on a direct intuition of the order in space and time. "In other words, when faced with two drawings representing different pairs of levels, the child cannot decide with certainty which of the pairs precedes the other; this is because instead of directly perceiving the movement of the water from top to bottom or vice versa, they see nothing but static spatial relationships (or static levels) to which an order must now be given; they must reconstitute this order by deduction, in the form of a temporal succession." (Piaget, *ibid.*, p. 12.)

We are particularly interested in the next stage. The child is now capable of arranging the drawings in order, although he is still groping. But his success is still due to intuition. If the drawings are cut in half, so that the top bottles are separated from the bottom ones, the child will not be able to put the right bottles together for every level, or to put it differently, he will not be able to establish the coseriation of the successive levels in each of them. He is capable of a general reconstitution when the drawings are not cut in half, but he does not know how to deduce the level in the lower bottle from that in the upper bottle, or vice versa.

Therefore, between complete failure to reconstitute an order and success based on operational constructions, there is an intermediate stage in which the child is able to form a representation, although still through intuition, of the sequence of positions, by a general comprehension of the movement as a whole. In the experiments involving the movement of two figures in the same direction, the progress of the child is marked by his ability to dissociate the temporal order of their arrivals from the spatial order; to do this he must be able to form a *representation* of one

independently of the other. This dissociation only gradually becomes possible because the spatial order is perceived for longer
and is therefore more obvious than the temporal order. But before the child can pass from this intuitive stage to a stage where
his replies are based on reasoning, he must be capable of taking
the order of departure, the order of arrival and the duration of
movement of each figure into consideration simultaneously. At
the stage of simple representations, the child has still not learned
to use the durations as a basis for reasoning any more than causal
deductions. Even children who are capable of forming correct
representations of the durations can deduce nothing from these
to indicate the order of arrival; inversely, children who are capable of forming a representation of this order do not draw any
exact conclusions from this to help with the durations (Piaget,
ibid., pp. 94-99).[3]

To sum up, the progress of a child toward temporal seriation
of events is marked by the replacement of the stage of perception
by the possibility of constructions on the level of representation,
although they still remain intuitive. In the next stage he actually
understands the order because he is capable of using all the facts
of the situation—acquired knowledge, causal relationships, and
in particular the durations between events—in an operational
construction.

The reason for this progress is obvious. As the child's intelligence develops, he becomes aware of the mistakes in the replies
which satisfied him a few months or years earlier. This consciousness grows as a result of his failures to adapt and especially of
his gradual understanding of the relationship between the various
elements of his experience, which seemed to contradict each other
on the plane of perception or intuition, or at least to lack co-
herence. This progress in the serial reconstruction of events
through the increasing plasticity of his representations is shown
in his most commonplace behavior. Young children cannot reply

[3] The part played by the relationship of order and duration is only fully
appreciated when there are several series of temporal events interlaced.
When there is only one series of similar data (positions of a moving figure,
succession of notes in a melody, etc.), the relationship between order and
duration is one of simple interaction. In a succession A–B–C, the interval
A–C is greater than A–B, but this relationship is more or less intuitive.

to the simplest questions regarding the date: the day of the week, the month, the year. Their first accurate replies are based on undigested knowledge: "They have been told" . . . "it is so," but gradually their replies come to be guided by the sequential order of the days or the months and they rediscover or confirm the date by means of a mental construction founded on other dates which are used as cues. This process is not frequently used until the child is at least 5, even if he has a high IQ (Farrell, 1953).

b. Duration. The child's progress as regards duration follows the same pattern. The young child's apprehension of duration relates to its content; it is proportional to the work accomplished or to the changes perceived during it. The first sign of progress will be his acquisition of the ability to form a representation of the duration as an interval independent of what occurs in it. In other words, durations, which are at first dissimilar, will become more and more alike as he sees that they are common to all events, whatever the nature of the latter. What are the stages of this development? First, it can only result from the continual questioning of an experience which does not change with age. When we form a representation of the duration as an interval independent of what happens in it, this always implies that we doubt our immediate experience, whereas progress in the apprehension of the order of phenomena is, as we have seen, a passage from confused experience to a clearer experience. This evolution toward the representation and conception of a homogeneous duration is a slow process which we should like to follow from its very birth.

Here again we shall be much indebted to Piaget. We shall quote his work freely, viewing the facts he established from our own perspective and subsequently discussing his interpretation.

The homogeneity of time is apprehended gradually, in our opinion, as the child comes to realize more and more that his intuitive appreciations of duration are contradicted either by other intuitive appreciations or by estimations based on other cues. In other words, this homogeneity is born of the discordance between various modalities of his appreciations or between his personal appreciation and that of other people. The ability to conceive of a homogeneous and uniform flow of time implies "a liberation and decentration of thought with regard to the experienced duration." (Piaget, *ibid.,* p. 51.)

The child's faith in his first judgment begins to be shaken when he realizes that there are several possible estimations of duration. We shall use an experiment already quoted in Chapter 7 (p. 244). Water flows from a flask down a Y-shaped tube with two identical branches, into two different vessels. The two branches are regulated by a single cock; the volume of water flowing into the two vessels is therefore equal, and stops and starts at the same time. The shape and size of the vessels are different, so that when one is full the other is only one third or one half full. Young children fail to see that the water flowed for the same length of time into the two vessels, because the work accomplished seems to be different. They all agree that the water flowed for longer into the smaller bottle, because it is full. As regards the perception of order, their answers show that they do not grasp the simultaneous cessation of flow; they also do not think that the same quantity of water has flowed into the two vessels but believe that there is more in the full one. But an older child soon sees that the two flows stop and start together; he therefore admits that the water flowed down both tubes for the same length of time, but this knowledge is still intuitive and is thus still contradicted by the first method of appreciation. The following dialogue shows this coexistence of two contradictory conclusions:

Pat (6 years, 4 months) has understood that more time is necessary for filling the large vessel G than the smaller vessel C, and he has recognized the fact that the two flows of water stop simultaneously, but if he is asked: "Did it take the same length of time (for C to fill and G to be ⅓ filled)? *No. This one* (G) *took less time because it is not completely full.* How much time? *One minute for C, and less for G because there is not much in it and it is bigger.* So it took longer for one than for the other? *Oh, the same time because they were filled at the same time.* Why the same time? *Because that one* (C) *is small and that one* (G) *is big, but it wasn't completely filled."* (*Ibid.,* p. 139.) In this conversation it is the relationship of order which leads the child to change his conclusion, but this conclusion is still intuitive.

Other children continue to believe that the two durations of flow are unequal but agree that there is the same quantity of water in each vessel, despite the difference in shape, and conclude from this that the durations are equal. These children reconsider their first impression after an objective evaluation of the work accomplished according to the quantity of water which has flowed.

This juxtaposition of impressions associated with the influence of the order or with various simultaneous appreciations of the quantity of changes is to be found in other situations as well.

In the experiment with the apple and pear trees, for instance, some children may be seen to hesitate, when determining the respective ages of the trees, between a judgment based on their development and one based on their actual age, that is, on the time since they were planted.

The comparison brought about by this hesitation is clearly seen in an older child in the experiment where he has to compare the (objectively equal) durations of transferring pieces of lead and pieces of wood, the former being more difficult to move. Pim (10 years, 8 months) replies: *"It must be the same time. I was going to say that I took longer with the bits of lead, but then I thought that it must be the same thing.* Why? *Because I moved more bits of wood than lead."* (*Ibid.*, p. 256.) This child first estimated the duration according to the number of changes experienced. The pieces of lead attract more attention to themselves than the pieces of wood. But at the same time he saw that he had moved more pieces of wood, and this fact made him doubt his first impression. He judges the duration by using simultaneously both the criteria which may guide younger children: the work accomplished and the changes felt. These lead to two contradictory judgments, and he wisely concludes that the two durations must have been equal.

Younger children use only one criterion and stick to it. The possibility of seeing more than one point of view is an essential part of the development of his intelligence. It shows the relative value of the criteria used. When a child doubts his immediate impressions, he is brought of necessity to attempt an evaluation of the duration in itself and to picture it independently of its content. In this process a large part must be played by the child's social background and his progressive use of clocks and watches. Before the child is even capable of comparing several personal evaluations of duration, these are contradicted by his surroundings. If he has found the time very short while at play he is reminded by "the grownups" that the time for some disagreeable task has already arrived, and if he has found the duration of such a task to be long, he is reminded that he has not spent long enough on it. As he learns to use clocks, he finds that they confirm the adults' point of view.

At this stage, the child does begin to believe in the homogeneity of time as measured by clocks. When he was younger he did not; when his speed of work changed he thought that the speed of flow of the sandglass or the speed of the stop watch hand had changed; when he worked faster they also seemed to go faster (effect of assimilation) or more slowly (effect of contrast). But as the child compares more and more often the number of changes he feels or observes with those of the clock, he comes to the conclusion fairly quickly that the clock always moves at the same speed and is therefore independent of other external changes. After having worked once quickly and once slowly, but for the same duration and watching the sandglass, May (aged 6 years, 6 months) is asked: "Did the sand flow at the same speed or more quickly or more slowly? *More quickly . . . No, the same. . . . No.* The same or more quickly? *The same.* Why did you think it went more quickly? *You just think it does, but it's because you are going faster.*" (Piaget, *ibid.,* p. 187.)

It becomes possible for the child to see that the two durations are identical because he gradually dissociates the experience of his work from the experience of the movement of the sand or the hand of the clock, and because he sees intuitively that equivalent changes take place in equivalent periods. Historically, this is how man gradually came to have the conception of homogeneous time. The equality of two successive periods cannot in fact be measured directly but only by the equality of other periods; it therefore remains a supposition although it is verified all the time by the concordance of observations and the measurements it permits.

At the stage reached by the child we are discussing, isochronism, which is connected with a direct experience of the homogeneity of the changes in a given series, does not yet permit of the measurement of duration, for this implies the retention of one idea of time when we pass from one change to another. At this age 30 seconds by the sandglass are not the same as 30 seconds on a stop watch (Piaget, *ibid.*, pp. 191-196). The hand moves faster than the sand and the child concludes that he can do more during the time of the watch than during the time of the sandglass. But he is nevertheless capable of recognizing the fact that the watch and the sandglass start and stop at the same time and have therefore worked for the same length of time. He therefore does not infer the measurement of movement straight away from the equality of these durations; but by comparing the latter with each other and with the heterogeneous changes that fill them, he will probably understand that a duration can be independent of its contents. This representation of homogeneous time is actually a slow conquest and we shall see later that it is not finished even at adolescence.

2. The Relating of Order and Duration

When the child is about 7–8, there seems to be a fairly sudden reorganization of the data concerning order and duration (Piaget, *ibid.*, pp. 277-278); the child becomes capable of passing from one intuition to the other in a reversible construction which implies an understanding of the relationships between the seriation of events and their durations.

Generally speaking, the child acquires the ability to use spatial and kinesthetic information as a basis and explanation for his temporal intuitions. In the experiment where water flows in several stages from one flask into another, a child who has reached the operational stage can reconstitute the order of the levels of the liquid in the two vessels simultaneously and thus effect a coseriation. He is also able to see that the duration in which the upper flask empties is the same as that during which the lower one fills; he deduces this either from the fact that there is only one flow of liquid (simultaneity of starting and stopping) or from the fact that the same quantity of water is involved.

At this stage in his development, equilibrium is established between the intuitions of order and duration. "Although children see the relationships of succession and those of duration at the beginning of the stage of heterogeneous intuitions, without being aware of their necessary relationship with each other, these end by determining each other in a single comprehensive system which is at the same time differentiated and entirely coherent." (Piaget, *ibid.*, p. 75.)

Through this reciprocal relationship the child can relate series of changes which are independent of each other and understand their temporal relationships from the point of view of both the succession of events and the intervals of duration, without being misled by impressions bound up with the nature of the changes themselves. He can thus take into account all the temporal conditions of his experience because he understands and locates them; he can thus truly *construct* his notions of time through awareness of the relationships which exist between all the aspects of a change.

At this stage the child at last learns to measure time. The measurement of time, like that of space, implies that we observe the coincidence of the beginning and the end of an interval with those of another interval which is taken as a unit, or in other words that we grasp the synchronism of two different successions of change.

We shall disregard for the moment the problem we have already mentioned of the nature of the unit of time, which depends on the assumption of the isochronism of identical periods of change. The validity of this unit is not in question in the operation of measuring time. Any measurement implies the comparison of two durations. This is simple when the changes take place at the same time. As long as the child only has to determine whether one duration is longer or shorter than another, he only needs to estimate the durations according to the quantity of changes felt or the work accomplished. But the reply will not be accurate unless the changes being compared are identical. The comparison only becomes a true measurement when the reply does not depend on an intuitive appreciation of the durations but on their interrelationship, i.e., when the child takes the order of the beginnings and ends into account.

Judging the equality of two durations is a more complex process but it is indispensable for the actual measurement of time. In this case the comparison of intuitions is not enough except in the exceptional case of two changes being synchronous and of the same nature; for instance, when two objects are moving along parallel lines at the same speed. We have already seen this case in the experiments where a 5-year-old child had to compare the duration of two parallel movements taking place at the same time; 88 percent of the replies are correct when the speed is the same, the moments of arrival and departure simultaneous, but the movements in opposite directions. Spatial information cannot interfere, and the symmetry of the two movements leads to correct replies. But, as we have seen, if the two figures leave the same line at the same time, moving in the same direction, and A moves for the same length of time as B but at twice the speed, so that he covers twice the distance, only 17 percent of the children are capable of replying that the two durations of movement are the same. Their intuition of the changes gives them no help because A goes twice as fast and covers twice as much ground as B; the children are not yet capable of deducing the equality of duration from the simultaneity of departure and arrival (Fraisse and Vautrey, 1952). When the child is able to relate the two durations to each other by comparing the succession of the beginnings and ends, he can evaluate the relative length of any two durations and use one series of changes as a reference for the comparison of all durations. This is true measurement.

Apart from this operation, however, measurement also implies that the intervals between periodic changes are seen to be identical. We have seen that kinesthetic and spatial intuitions provide an intuitive basis for this observation; from this it does not necessarily follow that a 7- or 8-year-old child can already *conceive* that a clock measures homogeneous time independently of the changes which take place within it. The ability to measure time, which the child acquires through concrete operations, is apparent before his actual conception of time in general, or more exactly of what time, as measured by the clock, represents. We shall return to this point.

3. Piaget's Thesis

Before continuing with our study of the evolution of the notion of time in children, we must compare the ideas we have put forward with those suggested by Piaget, which we have referred to so frequently in the course of this chapter and Chapter 7. As we have already shown (Chapter 6, p. 154), he believes that a young child in the very first years of his life is able to realize subjective series by observing the results of his actions. In the years which follow, he relearns on the plane of intuitive thought what he already possessed on a completely practical level. At this stage, as soon as he is old enough to answer questions, we find that he *fails to differentiate between temporal and spatial order.* "Farther" means the same thing as "more time" because the child does not take the speed into account, or more exactly cannot see the inverse relationship of speed to time. The child judges the duration according to the content of the action, the quantity of work accomplished, or again by the external results of the action, of which the distance covered is only one particular instance (*Épistémologie génétique*, II, p. 27). At the stage of immediate intuition, time is therefore relative to the *result* of the action; it is a local time, peculiar to each movement and not uniform from one movement to another unless the speeds of change are identical. In this case the durations are proportional to the distance covered or, more generally speaking, to the changes felt or observed. There can, therefore, be no coordination of movements of different speed as this would entail the relating of space, time, and speed. "It is characteristic of the beginnings of thought to consider the perspectives of the moment as absolute and therefore not to group them according to their reciprocal relationships." (*Ibid.*, p. 275.)

These relationships appear in a later stage, that of *articulate intuition,* or the intuition of relationships. At this stage the child is no longer content to judge the duration according to the *results* of the action; he becomes capable of introspection and of an estimation of the duration *during* the action itself. As soon as he is capable of this introspection, it reveals an inverse relationship between the speed of the action and its duration. ". . . Older children find, as we do, that quick work seems shorter and slow work longer. *It is this introspective discovery*[4] which seems to be at the root of the inversion of the relationship between time and speed, for in the duration experienced during the action itself, time shrinks (for our consciousness) as a function of speed, whereas in the case of the evaluation of a duration in the memory, well-filled time expands and empty time is resorbed." (*Ibid.*, p. 29.)

The child is now capable of judging that the faster moving object takes less time; a dissociation takes place between the work accomplished and the activity itself, as this is apprehended introspectively. On the psychological plane, this activity is the equivalent of "power," or of strength multiplied by speed.

This stage is preparatory to that of true temporal operations. A child who has learned to distinguish between space and speed can realize the "coordination of these speeds, which will lead to a differentiation between the temporal order, the order of spatial succession and the durations of the distances covered" (*ibid.*, p. 30), and it will also show the relationship between the work accomplished and the activity felt. At this point, decentration in relation to time is accomplished and the child becomes capable of conceiving of a time which is homogeneous and reversible.

We shall leave this last stage, on which we reach the same conclusions, and return to the points on which the results of our experiments and considerations differ from Piaget's.

At the beginning there is no disagreement; a child's intuition of time is relative to its contents. There is no apprehension of duration on its own, without regard for what is enduring. But we have already stressed the fact—which we think is confirmed by our experiments—that young children judge duration not only according to the work accomplished, particularly the space covered, but also according to the activity involved or attributed to an object whose duration of movement is to be evaluated. In other words, a young child, like an adult, can judge duration either by the work objectively accomplished or by the changes he subjectively feels. According to the particular situation and his per-

---

[4] The italics are ours.

sonal attitude, he is more or less aware of the changes taking place around him, that is of the work accomplished, or of those he feels as he acts himself or identifies himself with the activity of another object, whether animate or inanimate.

Speed is not taken directly into account in these appreciations, but it can have some indirect effect. If we compare two movements of the same duration, more is obviously achieved in the quicker movement and therefore more changes are effected (quantity of work, distance, etc.). Quicker movement may also lead to the illusion that more speed = more time, because if we work intensely each "step" costs a greater effort. That speed is not directly involved is shown by the fact that the same illusion may occur in the opposite situation: it sometimes happens that every "step" takes more effort when we go slowly because the task is difficult. We have already quoted Piaget's observation that a walk in deep snow seems to take a long time.[5] Thus everything depends on the resistance to be overcome and this may arise from the effort to go as fast as possible or from fatigue or in general from the difficulty of the task. The immediate intuition of duration is that of an interval whose length depends on what happens in it. As in Oppel's illusion, this seems longer the more there is in it to attract our attention.

Some of Piaget's results actually point to the reverse. He based his reasoning mainly on the evaluation of the duration of physical movements and in this case the amount of work accomplished is of course predominant, for the information provided by the changes felt by the subject is important only when he is acting himself. In our experiments on the movements of figures, however, we found that the children gave equal importance to the criterion of the work accomplished and to that of the changes felt because some of them identified themselves with the figures representing racing cyclists. When the child has to evaluate the duration of his own action, Piaget himself noted that the stages are not so well defined; in the experiment with the pieces of wood and lead, he observed that some young children found the

---

[5] The influence of effort may be reinforced by that of effects of contrast between the actual duration and that we should like. But we do not always strive against the clock.

time spent transferring wood was longer (the durations always being objectively equal) because they moved more, but most thought they had spent longer transferring the lead because this work was more difficult. Piaget says that the children do not assume the equation "more difficult = slower = more time." He suggests instead that they still refer to one aspect of the work accomplished: "In the case of greater speed, the work is seen in terms of the distance covered . . . , in the case of the slower speed it is associated with the weight, the size, etc." (*Le développement de la notion de temps chez l'enfant*, p. 225). This interpretation seems very close to ours. The weight is judged as a function of the effort involved, which is what we have called the changes felt.

Thus we do not consider that the estimations based on introspective information are characteristic of the second stage, because they are also found during the first stage. It therefore does not seem likely to us that operational equilibrium is reached through the relating of the work accomplished to the speed of change (or of the activity when the time of an actual action is involved).

We think that between the stage of the intuition of order and duration and that of operations there is an intermediate stage. During this stage, intuition is gradually transformed into more and more abstract representations; the order of events is found by a construction which corresponds more and more exactly to reality as the child's experience is enriched. The representation of duration becomes more and more independent of what happens in it. There is therefore a progressive differentiation between the intuitions of distance, speed, and duration and these three factors can thus gradually be related to one another. At this phase of their development, children are capable of grasping, for instance, that the faster moving of two objects takes less time than the slower to cover a certain distance or accomplish a task. Thus the child is liberated, as it were, when he can relate order and duration and so he acquires the notion of time.

This interpretation is the same as that suggested by Piaget for the evolution of other notions, for he believes that in general a child passes from the stage of sensory-motor adaptation to that of a representation which develops gradually until it leads to reversibility at the point where he becomes capable of operations.

What accounts for this difference between Piaget's interpretation and ours? Perhaps we can explain it by the original orientation of Piaget's work on time. In his Foreword he says himself that Einstein had asked him whether the subjective intuition of time was "immediate or derived and whether it was integral with speed from the first or not." With his attention thus oriented, Piaget considered the problem of time with particular regard to its relationship with speed. One might almost say that he sought situations in which the relationship time $= \dfrac{\text{distance}}{\text{speed}}$ was apparent. He affirms several times that the *notion* of time only takes shape when there is first articulate intuition and then the relating of speed and duration. We, on the other hand, tend more toward the idea that the representation of a duration, being already of an abstract nature and constituting a ground on which several changes are located, appears before the point at which the child becomes capable of relating order and duration by also taking into account the logical connections between time, distance, and speed. We also do not believe that life, which gives the child occasion to meet the resistance of time, presents it to him as a relationship between the work accomplished and the speed of the action. According to Piaget, the child makes this "discovery" at the stage of articulate intuition, when he is capable of introspection and sees that, "while it is being experienced, a rapid or accelerated action brings about a contraction of time (by virtue of the inverse relationship of time to speed), whereas the duration expands in the memory (because it is then judged by the distance covered or the work accomplished)." (*Ibid.*, p. 266. See also the passage from *Épistémologie génétique* quoted on p. 273.)

Let us disregard the questions raised by the discrepancies between the immediate impression of experienced duration and that which we have in retrospect. Let us merely consider the fact that for Piaget, rapid action gives rise to a contraction of time. On this point we disagree with him; the apparent contraction and expansion of duration are not related to the speed of the action. We have already discussed this: sometimes speed increases the number of changes felt when it entails an effort and centration on the task (running, writing quickly, etc.), and sometimes it can decrease it when the action is rapid only because it is easy. In both cases we judge the duration not in relation to the speed but according to the effort made and the changes felt. There does not seem to us to be an inverse relationship of time to speed on the plane of intuition, but only on that of operations.

This discussion may seem a little unnecessary, but it does raise one essential question: do we eventually acquire an intuition of duration as such or is it always relative to space and speed? Piaget supports the second alternative, saying that ". . . the fundamental intuitions are those of distance and speed; time is gradually distinguished from these, but only in so far as the two movements can be related to each other. . . ." (*Ibid.*, p. 42.)

In our opinion, the young child has intuitions not only of speed and distance but also of duration. He sees the latter in the elementary form of an interval which stands between him and the fulfillment of his desires. This interval seems at first to consist of nothing but the number of changes which take place in it, but as the child develops, the intuition is transformed into a representation, an abstract ground, the scene of changes but independent of them. This is, of course, only one stage, for the child cannot take all the aspects of a change into account until he has learned to relate the order of events and the durations which separate them and to pass from one system of information to the other. At this final stage we find ourselves back in complete agreement with Piaget, that "operational time is reached when the order of successions can be deduced from the interrelation of the durations, and vice versa" (*ibid.*, p. 278), but we think that the *representation* of time exists independently and is apparent before the operational stage.

4. The Evolution of the Notion of Time up to Adolescence

When a child of 7 or 8 relates the changes he experiences to one another and succeeds in establishing coseriations, he does not yet realize that time is a relationship independent of change. He only gradually reaches this level of abstraction.

Some important research by Michaud (*Essai sur l'organisation de la connaissance entre 10 et 14 ans,* 1949) explains this evolution. He asked groups consisting of a large number of children between 10 and 15 years old the following question: What happens to time when we put the clock forward an hour in spring and jump suddenly from 11:00 P.M. to midnight? To put the problem in a more concrete form, he also asked them whether they suddenly became older. The answers to these questions show the relationship established spontaneously by the children between the time by the clock and the time of other changes, days, nights, or growth.

Disregarding those who did not understand the question, Michaud found that the children could be classed in four main categories according to their replies:

1. Those who regard time as a *real quantity.* When the clocks are put forward, time "vanishes" or "is taken away;" time itself is affected. The proof of this fact is that these children think they have grown older. "Yes, I grew older when we jumped from 11:00 to midnight; it was the same as if the whole hour had gone by" (13 years, 2 months). "An hour was added to our age; everyone grew older" (13 years, 11 months). Some of these children, however, reply that they did not grow older, but their remarks show that their conception of time is no different from that of the others in this group; they merely do not believe that anyone grows older so fast: "It didn't make me any older because you don't grow older in one hour" (13 years, 2 months).

2. Those children for whom putting forward the clock is purely a *practical problem.* The hands of the clock can be moved and this act makes us lose an hour "which could have been useful" (*ibid.,* p. 88). But it does not occur to them to wonder whether the loss of this hour has any relationship with an abstract time which is independent of the possibilities it opens for action.

3. The children who consider the jump of one hour as a *mathematical problem.* In winter the clock is put back and in the summer it is put forward; this means that $1 - 1 = 0$ (13 years, 1 month). This reasoning still does not mean that the child grasps the fact that other changes are not affected. The majority think that "I didn't grow older because *I didn't live the hour we lost*"

(14 years, 10 months); but some of these children think that their age is affected by the jump. "Yes, I suddenly grew an hour older because the clock was put forward an hour, but we get the hour back in winter and I get an hour younger then . . . so my age hasn't changed" (15 years). (*Ibid.*, p. 131.)

4. The children who recognize the fact that the time of the clock is a pure *convention* which does not affect the changes that occur in nature, particularly the movement of the sun and their own age. To move the hands is simply to create a *discrepancy*.

This last kind of reply is the only one which shows that the child *grasps* that the time of the clock is a convention which has no effect on the course of change; it is also the only reply which shows that this time is an underlying thread of uniform nature which is independent of human actions. Most of the replies in which the child considers the jump of an hour as a mathematical question also tend in the same direction. Such answers are given more and more often as the children are older.

If the different results obtained by Michaud are summed up in one table (*ibid.*, pp. 74, 89, 131, 241), it may be seen that the percentage of children (boys and girls) giving each type of reply follows a definite pattern with age.

Number of Children	Age	Time = Real Quantity, %	Time = Activity, %	Time = Mathematical Problem, %	Time = Convention, %
247	10	36.8	36	2	19.8
336	11	32.1	33.3	2	29.7
478	12	22.5	25.7	3.1	39.1
459	13	16.5	22.4	5	47.7
219	14	16.4	11.7	10.3	56.8
59	15	10.1	6.7	23.7	59.3

From this table it is apparent that at the age of 10, about ¾ of the children do not yet regard time as an abstraction and for them the change in time influences their age; they do not grasp the fact that the changes due to age are independent of the clock. Only at 13 do 50 percent of the children show that they have understood that the time of the clock is a convention which has no effect on the changes it measures.

These results complement our foregoing discussions. At the end of the first stage in his development, an 8-year-old child becomes capable of relating the durations of two series of changes; but when he does this he still ascribes concrete properties to time and time is still bound by these. Clocks give him the model of a continuous state of change with a constant speed, thanks to which he grasps the identical periods which are to serve as units; he can then determine the duration of changes in relation to this particular change. Only at the age described by Piaget as that of logical operations, i.e., adolescence, is the child capable of passing from the concrete homogeneity of the time of the clock to the abstract homogeneity of a duration which is the thread linking events without being dependent on them.

Adaptation to time is therefore a function of the development of the intelligence and of the operational level reached at each age by the individual. Consequently it is not surprising that several authors (see Chapter 6, p. 181) have found very high correlations between various general intelligence tests and the results obtained in a questionnaire relating to temporal orientation, the divisions of time and the ways of dating events, all knowledge which must be learned but which cannot be understood until the processes of the measurement of time have a meaning for the child.[6]

III *Representation and the Notion of Time*

To conclude this chapter devoted to the higher forms of adaptation to time in humans, we must consider the ultimate end of this genetic evolution, which we have described at such length, and define more precisely the actual nature of what man

[6] The significance of these correlations has been made clear by the research of Gothberg (1949) on cases of mental deficiency. The correlation between a questionnaire concerning the various concepts of time and the mental age is .84 but only .31 between it and the chronological age (the mental age being held constant).

calls time. Philosophers have debated this point without reaching the slightest degree of agreement. Obviously, as Nogué pointed out (1932), time is not a simple idea. Yet Pascal (*De l'esprit géométrique,* in *Pensées et opuscules,* p. 170) wrote that all men know what is meant by time even if they do not agree on its nature. Our various methods of adaptation to change give us a great number of experiences, but these give us only a partial explanation as long as they are not related to one another. It is this integration which we call time.

As we know, these experiences all boil down to two main ones, each of which has various aspects: *succession* and *duration.* As experiences, they *both* belong to the present.

The experience of succession is that of the evanescence of all our perceptions, and in particular of auditory perceptions.[7] More generally, when we have reached an age where we can distinguish between our perception and its object, our experience is that of the perpetual renewal of our sensations, our thoughts, and our emotions. Its principal characteristic is irreversibility, the impossibility of returning to the plane of the experience which was. But, thanks to memory, the consequence of this experience of succession is the development of temporal perspectives composed at the same time of the memory of past presents and of the anticipation of future presents, based on what has already been lived.

The experience of duration is that of an interval. It already exists in the perception of a succession in which the interval between successive stimuli is as much a reality as their order and number. We are, however, made most conscious of duration by the resistance it offers to the realization of our wishes and by our inability to make the desired object present at will. Later in the course of genetic development, duration is also revealed by the distance of our memories; the quantity of memories between two points in time is at the very root of our experience of the dimension of duration.

---

[7] "The most naked experience of time we can have is that of a birth and a death. B is born, starting from nothing, and A has just died. It is this connection between birth and death which I shall call 'becoming'." (Berger, 1950, p. 94.)

These two experiences have different contents and neither actually constitutes an *experience of time*. But they combine in suggesting to us one symbolical image which results from the necessity of considering all these experienced changes together in order to differentiate between them and arrange them in sequence. This image is that of a space in which all events are located, showing at the same time their multiplicity and their varying degrees of proximity. Our temporal perspectives, born of the multiplicity of past and future experiences, cannot be the object of a representation unless we place events side by side in relation to each other. This transcription is natural, because temporal order often coincides with spatial order and distances correspond to durations of movement. In the mountains, where movement is usually by foot, if a peasant is asked "is it far?" he is just as likely to answer "an hour" as "2 or 3 miles."

It is easy to understand the spontaneous transposition of the temporal to the spatial if we remember with Wallon the actual nature of our mental representations: "Being the means of immediate adaptation to surrounding reality, our states of consciousness, our perceptions only take up those of our relationships with the outside world which concern our existence. But these states of consciousness must express the useful aspects of our relationships in clear terms, by a system of impressions and symbols which can provide us with sharp distinctions and well-defined landmarks. . . . This law of the greatest utility orients the evolution of the conscious mind. Those of our states which cannot give rise to distinct representations by themselves will fade behind the symbolism of another series seen in more manageable and definite terms." (*Le problème biologique de la conscience*, p. 326.) Our visual sensations are better defined and therefore give rise to spatial images which enter our field of consciousness because they are best fitted to give us a useful representation of the world in which we live.

Although the spatial distribution of changes gives us a practical means of forming a representation of them, Bergson is still right in saying that this image of spatialized time does not correspond to any immediate experience. But does it not arise from the need not to be shut up in the present experience but to represent—or make present—past or future changes with their dual aspects of order and duration? Thus we let the dynamic aspect of the experience of becoming slip away. Could it be otherwise? Ever since the time of Heraclitus, philosophers have tried to define experienced duration by comparing it to a stream. The image is attractive, but if we analyze it we find that it is not a true interpretation of our experience. It does show the evanescence of this experience, but only for an observer who stands on the banks of the stream, outside time. It is also inadequate in other ways. The water flows toward its future, but for the observer it disappears into his past. If we turn to the experience of someone caught in the current, the change is on the banks and the stream can no longer be compared with time (Merleau-Ponty, *Phénoménologie de la perception*, 1945, pp. 470–471). This metaphor also does not allow a representation of the different moments of time, especially if these correspond to several series of changes. It is difficult to conceptualize becoming.

The ability to form a simultaneous representation of several successive moments, by placing them side by side and separating them by intervals or durations, permits us to complete our rough images of becoming. Through this imagery the order of changes depends only on thought and escapes the domination of experience. It makes time reversible; we can pass from a later to an earlier event as well as vice versa.

This representation becomes increasingly detached from the first images which constituted it and tends to become nothing more than the idea of a uniform and continuous background. It ends in a conception in which time is assimilated to a Euclidean type of space in which everything remains. Berger (1950, p. 102) showed that this construction of a representation expressed our desire to escape from a perpetual becoming that ends in death. "Time is the revolt of man against death, whose inevitability is revealed to him by the present, against this flowing away not of time but of its contents, against the fact that nothing remains in his hands. . . ."

We are not easily able to dissociate this representation of time—like that of space—from ourselves. We locate this background that we have conceived in relation to our own bodies and we orient it according to our habits of thought. This has been directly verified. Guilford (1926) asked his students to make a drawing which would represent the past, the present, and the future. Ninety-one percent of their drawings consisted of a movement from left to right, this direction obviously being due to the influence of reading and writing, from which it takes on a special significance in the West. Anyone who wishes to represent various stages of a development or to tell a story in pictures, as in comics, always proceeds from left to right. Apart from this general direction, Guilford also found that 58 percent of the drawings showed the past below and the future above; 22.5 percent depicted the present on the crest of a convex wave . . . others represented a broken line. This variety points to the fact that the spatialized location of changes is a very personal matter. If this task of locating representations is given an even more concrete nature by asking the students, as we have done, to locate the child they were and the old man they will be, the results are much the same as Guilford's: 31 percent locate the child on the left and the old man on the right; 11 percent put the past behind and the future in front; 10 percent put the past below and the future above, and 13 percent show more complex locations in which two of the three directions of space are seen simultaneously. It is interesting to note that 35 percent do not locate these images and explain that they are content to visualize them successively; this corresponds to a more primitive method of representing change: the juxtaposition of successive images. This is how time is often symbolized: successive portraits of the same man, successive pages of a calendar, etc. This considerable variety of replies shows from another point of view what we have already found in our genetic studies, that our representations of time may be more or less abstract according to the contents to which they refer.

The representation of time by a schematic continuum directed in one dimension[8] is adequate when we are thinking of changes

_____

[8] The representation of time by a straight line predominates at the moment, but this image is to some extent dependent on cultural developments. Aristotle informs us that in his day "the current idea is that human affairs are a circle . . ." (*Physics*, book IV). Line or circle, the reasoning remains the same.

that form a homogeneous series, like the different ages of our life, the successive states of our body, or the succession of days, months, or years. But it is not enough when we have to *relate* heterogeneous series of events to each other. In this case we have to construct time and locate events in relation to each other, although these do not fall into a natural order, and also relate the durations between them. To take an example from Piaget, if I wish to consider at the same time all the temporal data of my life, for instance to locate the events of my family life and the political events of my country, I proceed by a series of constructions in which dates and durations have their part (the war lasted five years, my second child was born three years after the first, etc.). I rely on representations, but ultimately no representations are possible, for we cannot imagine simultaneously several intervals of time which overlap. But this construction of time does not lead to a notion which includes one class of objects, but to a single and comprehensive scheme which we attain at the end of a series of operations which constitute the temporal universe (Piaget, *Le développement de la notion de temps chez l'enfant,* p. 293).

This temporal scheme or this notion is therefore not the result of a simple abstraction brought about by a multiplicity of experiences, for the simple reason that temporal experiences are heterogeneous. As Lavelle puts it (*Du temps et de l'éternité*): "Only a relationship can give us a representation of an object which is not present" (p. 192). This is why time is also not a simple idea. Time is born from the very activity of the man who tries to reconstruct the changes in which he takes part. And we can only master what we have reconstructed. With the notion of time we arrive at the most complete adaptation of man to the successions which form the thread of his existence. Man thus has the impression that his conception of time is that of an absolute time best formulated by Newton.

The steady progress of science, which has led to the theory of first simple and then general relativity, was to show, however, that what was thought to be absolute time was still nothing but "local time," as it was called by Lorentz, or more exactly "individual time" according to Langevin. Seriation is not fixed nor duration homogeneous except in relation to one system of references whose different parts are static in relation to one another. When there are various systems of reference which move in relation to one another, there is no common time.

The problem was brought to a head by the question of simultaneity; as we have seen, this is a fundamental question from the point of view of both seriation and the measurement of duration. At first sight we agree that two events are simultaneous when we perceive them together. But we soon learn that when two perceptions are simultaneous, it does not necessarily mean that the two events to which they correspond are also simultaneous. It all depends on the position of the observers in relation to the sources of the phenomena and the speed of transmission of the messages. In a fixed system of references, however, we can infer the simultaneity or nonsimultaneity of two phenomena, whatever our position, by considering the distances, the speed of transmission and the interval measured between the reception of the two messages. Wherever they are positioned, all observers will come to the same conclusion. But this is not true if the two events take place in systems which are moving in relation to each other, for instance in two different stars. In the case of absolute time, we claim that an event taking place on the sun and one on the earth can be simultaneous, but this is an unfounded belief. We realize this as soon as we set about trying to verify this simultaneity, since the observed phenomena and the observers are moving in relation to each other. The theory of relativity has shown that we can only measure between the two events a space-time interval, for the two phenomena can only be measured in relation to each other. The interval is a spatio-temporal fact.

A consideration of the problem from the point of view of duration leads to the same conclusions. There is no homogeneous duration, and hence no possible unit of time, except in a system where there is relative stability of physical conditions. The pace of our clocks depends on the field of gravitation and its accelerations and decelerations. If two clocks were exposed to different fields in different planets, they would not measure the same time. And because there is no absolute simultaneity, it would not be possible to regulate them in relation to each other.

The theory of relativity calls for a new conception of time as of space. Unlike the notion of time that we have studied, this conception does not arise from the direct action of man on things, his adaptation to the conditions of his life, but from his scientific activity. It was while trying to understand certain physical paradoxes, particularly the negative results obtained by Michelson, that Einstein abandoned the hypothesis of absolute time and decided that there must be a time peculiar to each system of reference.

This effort to determine the space-time relationships of the universe may be considered as a new attempt at better adaptation of our knowledge to reality, but it does not affect everyday psychological life, which is not at its source. We therefore think that the time of relativity brings us beyond the bounds of the psychological problem of temporally organized behavior.

ı11ı11ı11ı11ı11ı11ı11ı11ı11ı11ı11ı11ı11ı11ı11ı11ı11ı11ı11ı11ı11ı11ı11ı11ı11ı11ı11

CONCLUSION – THE VALUE OF TIME

"TIME MAKES ME AND I MAKE TIME!"[1]

From the day of our birth until our death, our body develops and changes continuously under the action of time. The conditions of our existence also change unceasingly and we are modelled in a thousand different ways by them. We live in the rhythm of day and night. Our nervous centers register the duration which elapses between a gratification and the signal which preceded it. Every event we experience is given as it were a temporal sign by its concomitance with some habitual change.

Social life is the background par excellence of our adaptation to change; it refracts, as it were, the transformations of the world around us. Is it not true that bringing up children consists essentially in teaching them to adapt the cycle of their activities and desires to the rhythms of adults? It is the parents first who fix the time of getting up, going to bed, meals, playtime, and work. Later school, a job, the town add their own demands. It is through living with others that we suffer from postponements forced on the fulfillment of our desires. Those two forms of adaptation, expectation, and the precipitation of an action, are aggravated and increased in number by our social life. When

---

[1] We have borrowed this aphorism from M. Bonaparte (1939), who used it to show the difference between Bergson and Descartes.

we submit to time it means for all practical purposes that we accept the time of others.

This temporal pressure of society covers a range of degrees (Stoetzel, 1953). Generally speaking it is more intense when we are caught up in a more complex network of social relations. These differences may be shown by a simple but significant example: The proportion of people who wear a watch in any town increases as its population grows. Farmers are obviously less bound by a precise timetable than office staff or workmen. For every one of us there are variations in the temporal pressure forced on us by the general framework of our lives; it varies according to our environment at any given moment, the day of the week or the time of the year. There is one time for the office, another for the street, another for the home (Halbwachs, 1947), and there is also a time for weekdays and another for Sunday, one time for work and another for holidays. This variation is beneficial because it entails an alternation of tension and relaxation and the successions of pressure and alleviation contribute toward the establishment of a rhythm for the individual's life. We all know the relief brought by a weekly rest, the holidays, those breaks in the infernal rhythm of town life.

However precious these pauses may be, there are moments when we try to escape completely from the pressure of time. Sleep is the boundary enclosing us in our biological individuality. Daydreams also free us to a lesser degree from the demands of reality, and particularly of social time. In daydreams our wishes are fulfilled as they are formulated; but it is a dangerous freedom when it becomes a habit, for it can lead to insanity, which is precisely a break between the individual and society. Toxic intoxication and—on another plane—mystic ecstasy are also means to liberation from time; they introduce us to the euphoria of eternity (Bonaparte).

We must not be misled by these exceptional experiences. The security of a normal human being does not lie in liberation from time. Temporal pressure is constricting, but it is also the framework within which our personality is organized. When it is absent we are disoriented. There is nothing to bind the sequence of our activities; we are alone. From this confusion there arises not only a feeling of emptiness but also a vague fear; we are afraid of facing unarmed the compulsions that the socialization of behavior usually inhibits or makes us suppress. When some people fervently seek new occupations to "divert" themselves from their anxiety and others bind themselves to rigid timetables, their actions arise from the same need. Human equilibrium is too precarious to do without fixed positions in space and regular cues in time.

The harmonization of individual times is the result of a multitude of interactions through which they tend gradually toward the cooperation which is necessitated by social life. In this play of mutual adaptations, every individual plays his part with all the dynamism of his personality. The very way in which he submits to temporal pressure, or revolts against it, reveals what he is and what he would be.

Generally speaking, when we submit to the time of our group we choose security; to shake it off, to state our independence, is a form of aggressiveness. By being punctual we are sure of the approval of those responsible for the timetable, or at least we avoid misadventures. It may often be found that very precise individuals suffer from some feeling of insecurity. Others use punctuality as a sort of weapon; they flaunt it as a model to their equals or inferiors and use it to put them in the wrong (Adler, *The Neurotic Constitution*, 1916). Although he does not consider these rather neurotic types of behavior, Halbwachs (1947) notes that, since the degree of exactitude required is not the same for every group, "we relax and take our revenge in certain surroundings for the exactitude demanded of us in others."

Lack of punctuality has the same ambiguity of significance. An individual may arrive late on purpose through indifference to the requirements of society, through a somewhat aggressive desire for independence, to annoy those who await him, or to create an opportunity of excusing himself and thus be the center of attention. He may also postpone an activity in order to create a tension which will be a source of satisfaction[2] (Meerloo, 1948;

[2] This is the theory of psychoanalysts who consider preoccupation with time as a manifestation of the anal personality, since waiting or making someone else wait gives rise to the same erotic pleasure as retention.

Adler, *op. cit.;* Fenichel, *The psycho-analytic theory of neurosis,* 1945, p. 282).

These complex implications of our adaptation to change show how strong is the domination of time in every aspect of our lives. Although punctuated by the rhythms of the earth, the changes within us are fashioned by the rhythms of society.

The efforts of man to rid himself of this domination of time are not limited to forms of escapism. A higher form of liberation consists less in freeing oneself from time than in conquering it; but this is not possible unless we escape from change. This is the end toward which the efforts of constructive thought have striven. Pierre Janet noted that philosophers have a particular dislike for time and have made every effort to destroy it (*L'évolution de la mémoire et de la notion de temps,* 1928, p. 496). We need not go so far; it is enough to observe man's behavior at the stage of his first intellectual adaptations: this behavior shows that from the very beginning man seeks to dominate change.

As a defense against time, man has first and foremost his memory; this keeps past changes present, reconstitutes their order, and reveals their significance. Man forges the unity of his personality by giving himself a past. And humanity claims a past and a future in this image. Different societies multiply the evidence of bygone ages by accumulating archives, libraries, museums. They immortalize themselves patiently by writing their own history.

The movement to draw all one's past to oneself and immobilize what was change is complemented by the symmetrical effort to anticipate the future and adapt it in advance to one's desires. The forethought of man extends beyond the limits of an allotment of time; it sends scholars, scientists, and politicians in search of distant goals.

This vision which embraces the past and the future as well as the present does not exhaust our capacity for organizing change. The work of memory is complemented by thought, which relates every sequence of events to others; we learn to switch easily from order to duration, from before to afterward and from afterward to before. When this process is complete, man holds sway over time, or the law of change. Paradoxically, this law is in a way outside change itself. Although originating in the experience of time lived, time thought does not keep even its most outstanding characteristics. It is not an abstraction of change, but it coordinates multiple series of changes and thus gives them meaning.

We see the heterogeneous nature of our representations of time in relation to experienced reality when we claim to have reached the heart of change through metaphor. We observed this with Merleau-Ponty: when we imagine the flow of time we are always spectators standing on the bank, watching a river flow by which remains unknown to us. Change is transformed into an object.

This transformation is the symbol through which we affirm our domination of time. The turns of speech we use every day are revealing. We speak of time as of a thing which is at our disposal: "Have you got time?" We give it a value like money: "to gain or lose time." We even barter with it or make it a manifestation of our generosity: "I gave him a lot of my time."

However complete this control over time may be, however detached from conscious reality, it cannot blind us to the indomitable nature of the experience of change. At every moment time is given to us only to be snatched away again immediately. It is a factor of all progress and improvement, but it can also ruin and destroy. Man is born and dies, he progresses and regresses, civilizations grow up and disappear; this is the double face of history which marks personal time and the history of the world alike with an essential ambivalence (Marrou, *L'ambivalence du temps de l'histoire chez saint Augustin*, 1950). Chronos, the Greek god of time, sires and devours his children. Janus, whose wisdom embraces the past and the future, is depicted with two faces, one laughing and one gloomy.

This ambivalence explains some of the choices we make. Led by our temperament, our situation, our past, each of us looks toward what time can offer or to what it destroys. From this spring the attitudes which color our daily behavior. We have shown the difference in value ascribed by different individuals to the two sides of the temporal horizon, the past and the future. These attitudes not only guide our actions, they inspire our philosophies. Each new system has its own conception of the opposing forces exerted on us by time; according to the importance it attributes in change to creation or destruction, the ego, the world, and even God, do not have the same meaning. There are indeed very few philosophies which refuse an important role to time, just as there are few men who shut themselves off completely from the future. But for those which claim that everything was created in the beginning, time can do no more than render explicit an implicit reality. For evolutionists, on the other hand, time is the creative force behind all progress. When the various aspects of time are thus metaphysically justified by philosophy, they attain their highest value.

Philosophies are born from our attitudes to time, and they rationalize and give values to these. They define more or less explicitly the better ways of living time and bring others into disrepute. The destiny of man and the destiny of civilizations bear the brand of the price attached to time by philosophies and religions.

The psychologist observes this sublimation of the experience of time and he seeks to understand its motivation and its significance. On the scientific plane he refrains from expressing opinions as to the value of time. If the choice is inevitable in his personal life, he is fully aware of the factors that determine it. Psychology will not direct his choice but it teaches him that there is more merit in making it consciously.

BIBLIOGRAPHY

B I B L I O G R A P H Y

ABE, S. An experimental study on influence of pause in paired comparison of time-intervals, *Jap. J. Psychol.*, 1931, **6**, 867–884.

ABBE, M. The spatial effect upon the perception of time. *Jap. J. exp. Psychol.*, 1936, **3**, 1–52.

ABBE, M. The temporal effect upon the perception of space. *Jap. J. exp. Psychol.*, 1937, **4**, 83–93.

ADLER, A. *The neurotic constitution.* Trans. B. Glueck & J. E. Lind. New York: Moffat, Yard & Co., 1916.

ADRIAN, E. D., & BUYTENDIJK, F. J. J. Potential changes in the isolated brain stem of the goldfish. *J. Physiol.*, 1931, **71**, 121–135.

ADRIAN, E. D. Electrical activity of the nervous system. *Arch. Neurol.*, 1934, **32**, 1125–1134.

AGGAZZOTTI, A. Sul più piccolo intervallo di tempo percettibile nei processi psichici. *Arch. Fisiol.*, 1911, **9**, 523–574.

AMES, L. The development of the sense of time in the young child. *J. genet. Psychol.*, 1946, **68**, 97–125.

ANDERSON, A. C. Time discrimination in the white rat. *J. comp. Psychol.*, 1932, **13**, 27–55.

ANDERSON, J. C., & WHITELY, P. L. The influence of two different interpolations upon time estimation. *J. gen. Psychol.*, 1930, **4**, 391–401.

297

ANDERSON, S. F. The absolute impression of temporal intervals. *Psychol. Bull.*, 1936, 33, 794–795.

ARISTOTLE. *Physics (Physique).* Trans. Carteron. Paris: Les Belles-Lettres, 1926.

ARNDT, W. Abschliessende Versuche zur Frage des Zahlvermögens der Haustaube. *Z. Tierpsychol.*, 1939–40, 3, 88–142.

ASCHOFF, J. Aktivitätsperiodik bei Gimpeln unter natürlichen und künstlichen Belichtungsverhältnissen. *Z. vergl. Physiol.*, 1953, 35, 159–166.

AUGUSTINE (Saint). *Confessions (Les Confessions).* Trans. Paul Janet. Paris: Charpentier, 1857.

AXEL, R. Estimation of time. *Arch. of Psychol.*, 1924, 12, No. 74.

BACHELARD, G. *L'intuition de l'instant.* Paris: Stock, 1932.

BACHELARD, G. *La dialectique de la durée.* Paris: Boivin, 1936.

BACHELARD, G. La continuité et la multiplicité temporelles. *Bull. Soc. franç. Phil.*, 1937, 37, 53–81.

BALD, L., BERRIEN, F. K., PRICE, J. B., & SPRAGUE, R. O. Errors in perceiving the temporal order of auditory and visual stimuli. *J. appl. Psychol.*, 1942, 26, 382–388.

BARACH, A. L., & KAGAN, J. Disorders of mental functioning produced by varying the oxygen tension of the atmosphere. *Psychosom. Med.*, 1940, 2, 53–67.

BARD, L. Les bases physiologiques de la perception du temps. *J. Psychol. norm. path.*, 1922, 19, 119–146.

BARNDT, R. J., & JOHNSON, D. M. Time orientation in delinquents. *J. abnorm. soc. Psychol.*, 1955, 51, 343–345.

BARUK, H. *La désorganisation de la personnalité.* Paris: Presses Univ. de France, 1952.

BEKESY, G. VON. Über die Hörsamkeit der Ein- und Ausschwungvorgänge mit Berücksichtigung der Raumakustik. *Ann. Physik.*, 1933, 16, 844.

BELING, I. Über das Zeitgedächtnis der Bienen. *Z. vergl. Physiol.*, 1929, 9, 259–338.

BENDA, PH., & ORSINI, F. Étude expérimentale de l'estimation du temps sous LSD 25. *Ann. Medico-Psychol.*, 1959, 1–8.

BENFORD, F. Apparent time acceleration with age. *Science*, 1944, 99, 37.

BENUSSI, V. Zur experimentellen Analyse des Zeitvergleichs. *Arch. ges. Psychol.*, 1907, 9, 384–385.

BENUSSI, V. *Psychologie der Zeitauffassung.* Heidelberg: Winter, 1913.

BENUSSI, V. Versuche zur Analyse taktil erweckter Scheinbewegungen. *Arch. ges. Psychol.*, 1917, 36, 59–135.

BERGER, G. Approche phénoménologique du problème du temps. *Bull. Soc. franç. Phil.*, 1950, 44, 89–132.

BERGLER, E., & ROHEIM, G. Psychology of time perception. *Psychoanal. Quart.*, 1946, 15, 190–206.

BERGSON, H. *Essai sur les données immédiates de la conscience.* 19th ed., Paris: Alcan, 1920.

BERGSON, H. *Durée et simultanéité.* Paris: Alcan, 1922.

BERGSTROM, J. A. Effect of changes in time variables. *Amer. J. Psychol.*, 1907, 18, 206–238.

BERMAN, A. The relation of time estimation to satiation. *J. exp. Psychol.*, 1939, 25, 281–293.

BERNOT, L., & BLANCARD, R. *Nouville, un village français.* Paris: Institut d'Ethnologie, 1953.

BERNSTEIN, A. L. Temporal factors in the formation of conditioned eyelid reactions in human subjects. *J. gen. Psychol.*, 1934, 10, 173–197.

BETHE, A. Die biologischen Rhythmusphänomene als selbständige bzw. erzwungene Kippvorgänge betrachtet. *Pflüg. Arch. ges. Physiol.*, 1940, 244, 1–42.

BIEL, W. C., & WARRICK, M. J. Studies in perception of time delay. *Amer. Psychologist*, 1949, 4, 303.

BIEMEL, W. *Le concept du monde chez Heidegger.* Paris: Vrin, 1950.

BINET, A. Note sur l'appréciation du temps. *Arch. de Psychol.*, 1903, 2, 20–21.

BIRMAN, B. N. Essai clinico-physiologique de détermination des différents types d'activité nerveuse supérieure. *Cah. Med. Sov.*, No. 2, 1953, 123–134.

BJÖRKMAN, M., & HOLMKVIST, O. The time order error in the construction of a subjective time scale. *Scand. J. Psychol.*, 1960, 1, 7–13.

BLAKELY, W. The discrimination of short empty temporal intervals. Ph.D. dissertation, University of Illinois, 1933.

BONAPARTE, M. L'inconscient et le temps. *Rev. franç. psychanal.*, 1939, 11, 61–105.

BONAVENTURA, E. I problemi attuali della psicologia del tempo. *Arch. ital Psicol.*, 1928, **6**, 78–102.

BONAVENTURA, E. *Il problema psicologico del tempo.* Milan: Soc. an. Istituto editoriale scientifico, 1929.

BOND, N. B. The psychology of waking. *J. abnorm. soc. Psychol.*, 1929–30, **24**, 226–248.

BOREL, E. *L'espace et le temps.* Paris: Presses Univ. de France, 1949.

BORING, E. G. Temporal perception and operationism. *Amer. J. Psychol.*, 1936, **48**, 519–522.

BORING, L. D., & BORING, E. G. Temporal judgments after sleep. *Studies in Psychology, Titchener commemorative volume.* 1917, 255–279.

BOUMAN, L., & GRÜNBAUM, A. Eine Störung der Chronognosie und ihre Bedeutung im betreffenden Symptomenbild. *Mschr. Psychiatr. Neurol.*, 1929, **73**, 1–40.

BOURDON, B. La perception du temps. *Rev. Phil.*, 1907, 449–491.

BOUVIER, R. *La pensée d'E. Mach.* Paris: Librairie du Vélin d'Or, 1923.

BRADLEY, N. C. The growth of the knowledge of time in children of school age. *Brit. J. Psychol.*, 1947, **38**, 67–78.

BRAUNSCHMID, M. Zeitsinn bei Fischen. *Blätt. Aquarien Terrarienk.*, 1930, **41**, 232–233.

BRECHER, G. A. Die Momentgrenze im optischen Gebiet. *Z. Biol.*, 1937, **48**, 232–247.

BROMBERG, W. Marihuana intoxication. *Amer. J. Psychiat.*, 1934, **91**, 303–330.

BROWER, J. F., & BROWER, D. The relation between temporal judgment and social competence in the feeble minded. *Amer. J. ment. Def.*, 1947, **51**, 619–623.

BROWMAN, L. G. Artificial sixteen-hour day activity rhythms in the white rat. *Amer. J. Physiol.*, 1952, **168**, 694–697.

BROWN, J. F. The visual perception of velocity. *Psychol. Forsch.*, 1931a, **14**, 199–232.

BROWN, J. F. On time perception in visual movement fields. *Psychol. Forsch.*, 1931b, **14**, 233–248.

BROWN, J. F. The thresholds for visual movement. *Psychol. Forsch.*, 1931c, **14**, 249–268.

BROWN, J. S. A note on a temporal gradient of reinforcement. *J. exp. Psychol.*, 1939, **25**, 221–227.

BRUSH, E. N. Observations on the temporal judgment during sleep. *Amer. J. Psychol.*, 1930, 42, 408–411.

BUCK, J. N. The time appreciation test. *J. appl. Psychol.*, 1946, 30, 388–398.

BÜNNING, E. Zur Kenntnis der erblichen Tagesperiodizität bei Phaseolus. *Jahrb. Bot.*, 1935, 81, 411–418.

BURCKARD, E., & KAYSER, CH. L'inversion du rythme nycthéméral de la température chez l'homme. *C. R. Soc. Biol.*, 1947, 141, 1265–1268.

BURGELIN, P. *L'homme et le temps.* Paris: Aubier, 1945.

BURTON, A. A further study of the relation of time estimation to monotony. *J. appl. Psychol.*, 1943, 27, 350–359.

BUYTENDIJK, F. J. J., FISCHEL, W., & TER LAAG, P. B. Über den Zeitsinn der Tiere. *Arch. neerl. Physiol.*, 1935, 20, 123–154.

BUYTENDIJK, F. J. J., & MEESTERS, A. Duration and course of the auditory sensation. *Comm. Pontif. Acad. Sci.*, 1942, 6, 557–576.

CALABRESI, R. *La determinazione del presente psichico.* Florence: R. Bempored, 1930.

CARREL, A. Physiological time. *Science*, 1931, 74, 618–621.

CATTELL, J. Psychometrische Untersuchungen. *Phil. Stud.*, 1885, 3, 1–72.

CHU TSI-TSIAO. Switching of short delay into long delay conditioned reflexes. *Pavlov. J. high nerv. Activ.*, 1959, 9, 512–519.

CLAUSEN, J. An evaluation of experimental methods of time judgment. *J. exp. Psychol.*, 1950, 40, 756–761.

CLAUSER, G. *Die Kopfuhr. Das automatische Erwachen.* Stuttgart: F. Enke, 1954.

CLIFFORD, W., & SCOTT, M. Some psycho-dynamic aspects of disturbed perception of time. *Brit. J. med. Psychol.*, 1948, 21, 11–120.

COHEN, J. Disturbances in time discrimination in organic brain disease. *J. nerv. ment. Dis.*, 1950, 112, 121–129.

COHEN, J. The concept of goal gradients: a review of its present status. *J. gen. Psychol.*, 1953, 49, 303–308.

COHEN, J. The experience of time. *Acta Psychol.*, 1954, 10, 207–219.

COHEN, J., HANSEL, C. E. M., & SYLVESTER, J. D. A new phenomenon in time judgment. *Nature;* London, 1953, 172, 901–903.

COHEN, J., HANSEL, C. E. M., & SYLVESTER, J. D. An experimental study of comparative judgments of time. *Brit. J. Psychol.*, 1954a, 45, 108–114.

COHEN, J., HANSEL, C. E. M., & SYLVESTER, J. D. Interdependence of temporal and auditory judgments. *Nature;* London, 1954b, **174**, 642–646.

COHEN, J., HANSEL, C. E. M., & SYLVESTER, J. D. Interdependence in judgments of space, time and movement. *Acta Psychol.*, 1955, **11**, 360–372.

COHEN, L. H., & ROCHLIN, G. N. Loss of temporal localization as a manifestation of disturbed self awareness. *Amer. J. Psychiat.*, 1938, **95**, 87, 95.

COLEGROVE, F. W. The time required for recognition. *Amer. J. Psychol.*, 1898, **10**, 286–292.

CONDILLAC, E. B. DE. Traité des sensations. In *Œuvres de Condillac.* Vol. III. Paris: Impr. C. Houel, 1798.

CONDILLAC, E. B. DE. De l'art de penser. In *Œuvres de Condillac.* Vol. V. Paris: Impr. C. Houel, 1798.

COOPER, L. F. Time distortion in hypnosis. *Bull. Georgetown Univ., Med. Centre,* 1948, **1**, 214–221.

COOPER, L. F. Time distortion in hypnosis with a semantic interpretation of the mechanism of certain hypnotically induced phenomena. *J. Psychol.*, 1952, **34**, 257–284.

COOPER, L. F., & TUTHILL, C. E. Time distortion in hypnosis and motor learning. *J. Psychol.*, 1952, **34**, 67–76.

COOPER, L. F., & ERICKSON, M. H. *Time distortion in hypnosis.* Baltimore: William and Wilkins, 1954.

COWLES, J. T., & FINAN, J. L. An improved method for establishing temporal discrimination in white rats. *J. Psychol.*, 1941, **11**, 335–342.

CURTIS, J. N. Duration and the temporal judgment. *Amer. J. Psychol.*, 1916, **27**, 1–46.

CZCHURA, W. S. The generalization of temporal stimulus patterns on the time continuum. *J. comp. Psychol.*, 1943, **36**, 79–90.

CZERMAK, J. N. Ideen zu einer Lehre vom Zeitsinn. *Wien. akadem. Sitzungsb.*, 1857, Nat. Cl., **24**, 231.

DAVIDSON, G. M. A syndrome of time agnosia. *J. nerv. ment. Dis.*, 1941, **94**, 336–343.

DECROLY, O., & DEGAND, J. Observations relatives au développement de la notion du temps chez une petite fille. *Arch. de Psychol.*, 1913, **13**, 113–161.

DE GREEFF, E. La personnalité du débile mental. *J. Psychol. norm. path.*, 1927, 400–454.

DELACROIX, H. La conscience du temps. In G. DUMAS, *Nouveau traité de psychologie*. Vol. V, pp. 305–324. Paris: Alcan, 1936.

DELAY, J. *Les dissolutions de la mémoire*. Paris: Presses Univ. de France, 1942.

DELAY, J., & BRION, S. Syndrome de Korsakoff et corps mamillaires. *Encéphale*, 1954, 43, 193–200.

DESCARTES, R. *Œuvres*. Ed. V. Cousin. Paris: F.-G. Levrault, 1825.

DEWOLFE, R. K. S., & DUNCAN, C. P. Time estimation as a function of level of behavior of successive tasks. *J. exp. Psychol.*, 1959, 58, 153–158.

DIETZE, G. Untersuchungen über den Umfang des Bewusstseins bei regelmässig aufeinander folgenden Schalleindrücken. *Phil. Stud.*, 1885, 2, 362–393.

DMITRIEV, A. S., & KOCHIGINA, A. The importance of time as stimulus of conditioned reflex activity. *Psychol. Bull.*, 1959, 56, 106–132.

DOBRZANSKI, J. Badania nad zmyslem czasu u mrówek. *Folia Biologica*, 1956, 4, 385–397.

DOBSON, W. R. An investigation of various factors involved in time perception as manifested by different nosological groups. *J. gen. Psychol.*, 1954, 50, 277–298.

DOEHRING, D. G. Accuracy and consistency by four methods of reproduction. *Amer. J. Psychol.*, 1961, 74, 27–35.

DOOB, L. W. *Becoming more civilized. A psychological exploration.* New Haven: Yale University Press, 1960.

DOOLEY, L. The concept of time in defense of ego integrity. *Psychiatry*, 1941, 4, 13–23.

DUDYCHA, G. J. An objective study of punctuality in relation to personality and achievement. *Arch. of Psychol.*, 1936, 29, No. 204. 53 pp.

DUDYCHA, G. J., & DUDYCHA, M. M. The estimation of performance time in simple tasks. *J. appl. Psychol.*, 1938, 22, 79–86.

DUNLAP, K. The shortest perceptible time interval between two flashes of light. *Psychol. Rev.*, 1915, 22, 226–250.

DUNLAP, K. Time and rhythm, *Psychol. Bull.*, 1916, 13, 206–207.

DURUP, G., & FESSARD, A. Le seuil de perception de durée dans l'excitation visuelle. *Année psychol.*, 1930, 31, 52–62.

EDGELL, B. On time judgments. *Amer. J. Psychol.*, 1903, 14, 418–438.

EHRENWALD, H. Versuche zur Zeitauffassung des Unbewussten. *Arch. ges. Psychol.*, 1923, 45, 144–156.

EHRENWALD, H. Gibt es einen Zeitsinn? Ein Beitrag zur Psychologie und Hirnpathologie der Zeitauffassung. *Klin. Wschr.*, 1931a, 10(32), 1481–1484.

EHRENWALD, H. Störung der Zeitauffassung der räumlichen Orientierung, des Zeichnens und Rechnens bei einem Hirnverletzten. *Z. ges. Neurol. Psychiat.*, 1931b, 132, 518–569.

EHRENWALD, H. Über den Zeitsinn und die gnostische Störung der Zeitauffassung beim Korsakow. *Z. ges. Neurol. Psychiat.*, 1931c, No. 134, 512–521.

EHRENWALD, H. Zur Hirnlokalisation von Störungen der Zeitauffassung. *Arch. Psychiat.*, 1932, 97, 683–684.

EISSLER, K. R. Time experience and the mechanism of isolation. *Psychoanal. Rev.*, 1952, 39, 1–22.

EJNER, M. *Experimentelle Studien über den Zeitsinn.* Dorpat, 1889.

EKMAN, G., & FRANKENHAEUSER, M. Subjective time scales. *Rep. Psychol. Lab. Univ. Stockholm*, 1957, No. 49.

ELKINE, D. De l'orientation de l'enfant d'âge scolaire dans les relations temporelles. *J. Psychol. norm. path.*, 1928, 28, 425–429.

ELLIS, L. M., et al. Time orientation and social class: an experimental supplement. *J. abnorm. soc. Psychol.*, 1955, 51, 146–147.

ESON, M. E., & KAFKA, J. S. Diagnostic implications of a study in time perception. *J. gen. Psychol.*, 1952, 46, 169–183.

ESTEL, V. Neue Versuche über den Zeitsinn. *Phil. Stud.*, 1885, 2, 37–65.

EYSENCK, H. J. *Les dimensions de la personnalité.* Paris: Presses Univ. de France, 1950.

FALK, J. L., & BINDRA, D. Judgment of time as a function of serial position and stress. *J. exp. Psychol.*, 1954, 47, 279–284.

FARBER, M. L. Suffering and time perspective of the prisoner. In *Authority and frustration. Univ. Iowa Stud. Child welfare*, 1944, 20, 153–227.

FARBER, M. L. Time perspective and feeling tone: a study in the perception of the days. *J. Psychol.*, 1953, 35, 253–259.

FARRELL, M. Understanding of time relations of five-, six-, and seven-year-old children of high I.Q. *J. educ. Res.*, 1953, 46, 587–594.

FAVILLI, M. La percezione del tempo nell' ebbrezza mescalinica. *Rass. Stud. Psichiat.*, 1937, 26, 455–462.

FEILGENHAUER, R. Untersuchungen über die Geschwindigkeit der Aufmerksamkeitswanderung. *Arch. ges. Psychol.*, 1912, **25**, 350–416.

FENICHEL, O. *The Psycho-analytic theory of neurosis.* New York: Norton, 1945.

FEOKRITOFF, Y. P. *Time as a conditioned stimulus to the salivary gland.* (In Russian.) Thesis, St. Petersburg, 1912.

FERRARI, G. C. La psicologia degli scampati al terremoto di Messina. *Riv. Psicol.*, 1909, **5**, 89–106.

FESSARD, A. Les rythmes nerveux et les oscillations de relaxation. *Année psychol.*, 1931, **32**, 49–117.

FESSARD, A. *Recherches sur l'activité rythmique des nerfs isolés.* Paris: Hermann, 1936.

FILER, R. J., & MEALS, D. W. The effect of motivating conditions on the estimation of time. *J. exp. Psychol.*, 1949, **39**, 327–337.

FINAN, J. L. Effects of frontal lobe lesion on temporally organized behavior in monkeys. *J. Neurophysiol.*, 1939, **2**, 208–226.

FINAN, J. L. Delayed response with pre-delay reinforcement in monkeys after removal of the frontal lobes. *Amer. J. Psychol.*, 1942, **55**, 202–214.

FISCHER, F. Zeitstruktur und Schizophrenie. *Z. ges. Neurol. Psychiat.*, 1929, **121**, 544–575.

FISCHER, F. Raum-Zeit Struktur und Denkstörungen in der Schizophrenie. *Z. ges. Neurol. Psychiat.*, 1930, **124**, 241–256.

FISHER, S., & FISHER, R. L. Unconscious conception of parental figures as a factor influencing perception of time. *J. Pers.*, 1953, **21**, 496–505.

FOUCAULT, M. *Le rêve.* Paris: Alcan, 1906.

FRAISSE, P., & FRAISSE, R. Études sur la mémoire immédiate. I: L'appréhension des sons. *Année psychol.*, 1937, **38**, 48–85.

FRAISSE, P. Études sur la mémoire immédiate. III: L'influence de la vitesse de présentation et de la place des éléments. La nature du présent psychologique. *Année psychol.*, 1944–45, **45–46**, 29–42.

FRAISSE, P. Recherches sur les lois de la perception des formes. *J. Psychol. norm. path.*, 1938, **35**, 415–423.

FRAISSE, P. De l'assimilation et de la distinction comme processus fondamentaux de la connaissance. In *Miscellanea Psychologia A. Michotte.* Louvain, 1947, 181–195.

FRAISSE, P. Étude comparée de la perception et de l'estimation de la durée chez les enfants et chez les adultes. *Enfance*, 1948a, 1, 199–211.

FRAISSE, P. Rythmes auditifs et rythmes visuels. *Année psychol.*, 1948b, 49, 21–42.

FRAISSE, P. Les erreurs constantes dans la reproduction de courts intervalles temporels. *Arch. de Psychol.*, 1948a, 32, 161–176.

FRAISSE, P., & OLÉRON, G. La perception de la durée d'un son d'intensité croissante. *Année psychol.*, 1950, 50, 327–343.

FRAISSE, P. La perception de la durée comme organisation du successif. *Année psychol.*, 1952a, 52, 39–46.

FRAISSE, P. Les conduites temporelles et leurs dissociations pathologiques. *Encéphale*, 1952b, 41, 122–142.

FRAISSE, P., & VAUTREY, P. La perception de l'espace, de la vitesse et du temps chez l'enfant de cinq ans. *Enfance*, 1952, 5, 1–20, 102–119.

FRAISSE, P., & MONTMOLLIN, G. DE. Sur la mémoire des films. *Rev. int. Film.*, 1952, 37–69.

FRAISSE, P., & JAMPOLSKY, M. Premières recherches sur l'induction rythmique des réactions psychogalvaniques et l'estimation de la durée. *Année psychol.*, 1952, 52, 363–381.

FRAISSE, P. La perception comme processus d'adaptation. L'évolution des recherches récentes. *Année psychol.*, 1953, 53, 443–461.

FRAISSE, P., & ORSINI, F. Étude expérimentale des conduites temporelles. I: L'attente. *Année psychol.*, 1955, 55, 27–39.

FRAISSE, P., EHRLICH, S., & VURPILLOT, E. Études de la centration perceptive par la méthode tachistoscopique. *Arch. de Psychol.*, 1956, 35, 193–214.

FRAISSE, P., & FLORÈS, C. Perception et fixation mnémonique. *Année psychol.*, 1956, 56, 1–11.

FRAISSE, P. *Les structures rythmiques*. Paris: Érasme, 1956.

FRAISSE, P. La période réfractaire psychologique. *Année psychol.*, 1957, 57, 315–328.

FRAISSE, P., & ORSINI, F. Étude des conduites temporelles. II: Étude génétique de l'attente. *Année psychol.*, 1957, 57, 359–365.

FRAISSE, P. L'adaptation du travailleur au temps. *Bull. Centre Ét. et Rech. psychol.*, 1958, 7, 79–83.

FRAISSE, P. Perception de la durée et durée de la perception. *Psychol. Franç.*, 1958, 3, 1–8.

FRAISSE, P., & ORSINI, F. Étude expérimentale des conduites temporelles. III: Étude génétique de l'estimation de la durée. *Année psychol.*, 1958, 58, 1–6.

FRAISSE, P. Of time and the worker. *Harvard Bus. Rev.*, 1959, 37, 121–125.

FRAISSE, P., & ORSINI, F. Étude des conduites temporelles. IV: La précipitation. *Psychol. Franç.*, 1959, 4, 117–126.

FRAISSE, P. L'influence de la fréquence sur l'estimation du temps. *Année psychol.*, 1961, 61, 325–339.

FRAISSE, P. et al. Comparaison des méthodes de reproduction, de production et d'estimation du temps. *Z. für Psychol.*, 1962, in press.

FRANÇOIS, M. Contribution à l'étude du sens du temps. La température interne comme facteur de variation de l'appréciation subjective des durées. *Année psychol.*, 1927, 28, 188–204.

FRANÇOIS, M. Influence de la température interne sur notre appréciation du temps. *C. R. Soc. Biol.*, 1928, 108, 201–203.

FRIEDMAN, K. C. Time concepts of junior and senior high school pupils and of adults. *Sch. Rev.*, 1944, 52, 233–238.

FRISCH, K. VON, & LINDAUER, M. Himmel und Erde in Konkurrenz bei der Orientierung der Bienen. *Naturwiss.*, 1954, 41, 245–253.

FRISCHEISEN-KÖHLER, I. Über die Empfindlichkeit für Schnelligkeitsunterschiede. *Psychol. Forsch.*, 1933a, 18, 286–290.

FRISCHEISEN-KÖHLER, I. Feststellung des weder langsamen noch schnellen (mittelmässigen) Tempos. *Psychol. Forsch.*, 1933b, 18, 291–298.

FROBENIUS, K. Über die zeitliche Orientierung im Schlaf und einige Aufwachphänomene. *Z. Psychol.*, 1927, 103, 100–110.

FRÖBES, J. *Lehrbuch der experimentellen Psychologie.* 2 vols. Freiburg-Br.: Herder, 1935.

FROLOV, J. P. Differentiation of conditioned trace stimuli and of trace inhibitions. *Arch. Biol. Sci.*, 1924, 24, 103–114.

FROLOV, J. P. Physiological analysis of the phenomenon of time registration in the central nervous system in animals and man. *15th int. Congress Physiol.*, Moscow, 1935, 102.

FULTON, J. F. *Physiology of the nervous system.* 3rd ed. London: Oxford University Press, 1949.

GAMBLE, F. W., & KEEBLE, F. The bionomics of convoluta roscoffensis. *Quart. J. microsc. sci.*, 1905, 67, 363.

GAMPER, E. Zur Frage der Polioencephalitis haemorrhagica der chronischen Alkoholiker. Anatomische Befunde beim alkoholischen Korsakow und ihre Beziehungen zum klinischen Bild. *Dtsch. Z. Nervenheilk.*, 1928, **102**, 122–129.

GARDNER, W. A. Influence of the thyroid gland on the consciousness of time. *Amer. J. Psychol.*, 1935, **47**, 698–701.

GASTAUT, H. L'activité électrique cérébrale en relation avec les grands problèmes psychologiques. *Année psychol.*, 1949, **51**, 61–86.

GENTRY, E. Methods of discrimination training in white rats. *J. comp. Psychol.*, 1934, **18**, 225–258.

GESELL, A. *The embryology of behavior.* New York: Harper, 1943.

GESELL, A., & ILG, F. L. *Infant and child in the culture of today.* New York: Harper, 1943.

GIEHM, G. Experimentell-psychologische Untersuchungen der Apperzeption des Zeitsinnes bei Geisteskranken. *Arch. Psychiat.*, 1931, **95**, 330–335.

GILLILAND, A. R. Some factors in estimating short time intervals. *J. exp. Psychol.*, 1940, **27**, 243–255.

GILLILAND, A. R., HOFELD, J., & ECKSTRAND, G. Studies in time perception. *Psychol. Bull.*, 1946, **43**, 162–173.

GILLILAND, A. R., & HUMPHREYS, D. W. Age, sex, method and interval as variables in time estimation. *J. genet. Psychol.*, 1943, **63**, 123–130.

GLASS, R. Kritisches und Experimentelles über den Zeitsinn. *Phil. Stud.*, 1887, **4**, 423.

GOLDSTONE, S., LHAMON, W. T., & BOARDMAN, W. K. The time sense: anchor effects and apparent direction. *J. Psychol.*, 1957, **44**, 145–153.

GOLDSTONE, S., BOARDMAN, W. K., & LHAMON, W. T. Intersensory comparisons of temporal judgments. *J. exp. Psychol.*, 1959, **57**, 243–248.

GOODFELLOW, L. D. An empirical comparison of audition, vision and touch in the discrimination of short intervals of time. *Amer. J. Psychol.*, 1934, **46**, 243–258.

GOTHBERG, L. C. The mentally defectve child's understanding of time. *Amer. J. ment. Def.*, 1949, **53**, 441–455.

GRABENSBERGER, W. Untersuchungen über das Zeitgedächtnis der Ameisen und Termiten. *Z. vergl. Physiol.*, 1933, **20**, 1–54.

GRABENSBERGER, W. Experimentelle Untersuchungen über das Zeitgedächtnis von Bienen und Wespen nach Verfütterung von Euchinin und Iodothyreoglobulin. *Z. vergl. Physiol.*, 1934, **20**, 338–342.

GREGG, L. W. Fractionation of temporal intervals. *J. exp. Psychol.*, 1951, 42, 307–312.

GREGOR, A. Beiträge zur Kenntnis der Gedächtnisstörung bei der Korsakoffschen Psychose. *Mschr. Psychiat. Neurol.*, 1907, 21, 19–46, 148–167.

GRIDLEY, P. F. The discrimination of short intervals of time by finger tip and by ear. *Amer. J. Psychol.*, 1932, 44, 18–43.

GRIMM, K. Der Einfluss der Zeitform auf die Wahrnehmung der Zeitdauer. *Z. Psychol.*, 1934, 132, 104–132.

GROETHUYSEN, B. De quelques aspects du temps. Notes pour une phénoménologie du récit. *Rech. phil.*, 1935–36, 5, 139–195.

GROOS, K. Zum Problem der unbewussten Zeitschätzung. *Z. Psychol. Physiol. Sinnesorg.*, 1896, 9, 321–330.

GROSS, A. Sense of time in dreams. *Psychoanal. Quart.*, 1949, 18, 466–470.

GROSSMUCK, A. Mit welcher Sicherheit wird der Zeitwert einer Sekunde erkannt? *Z. Sinnesphysiol.*, 1934, 65, 248–273.

GUILFORD, J. P. Spatial symbols in the apprehension of time. *Amer. J. Psychol.*, 1928, 37, 420–423.

GUILLAUME, P. *Introduction à la psychologie.* 3rd ed. Paris: Vrin, 1946.

GUINZBURG, R. L. È possibile l'appredimento di sensazioni eterogenee come perfettamente simultanee? *Arch. ital. Psicol.*, 1928, 6, 103–114.

GUITTON, J. *Le temps et l'eternité chez Plotin et saint Augustin.* Paris: Boivin, 1933.

GUITTON, J. *Justification du temps.* Paris: Presses Univ. de France, 1941.

GULLIKSEN, H. The influence of occupation upon the perception of time. *J. exp. Psychol.*, 1927, 10, 52–59.

GUNDLACH, R., ROTHSCHILD, D., & YOUNG, P. T. A test of analysis of set. *J. exp. Psychol.*, 1927, 10, 247–280.

GUYAU, J. M. *La genèse de l'idée de temps.* 2nd ed. Paris: Alcan, 1902.

HALBERSTADT, G. Notes sur les troubles de l'évaluation du temps chez les aliénés. *J. Psychol. norm. path.*, 1922, 19, 262–265.

HALBWACHS, M. La mémoire collective et le temps. *Cah. int. Sociol.*, 1947, 2, 3–31.

HALL, G. S., & JASTROW, J. Studies of rhythm. *Mind*, 1886, 11, 55–62.

HALL, W. W. The time sense. *J. ment. Sci.*, 1927, **73**, 421–428.

HARTON, J. J. The influence of the difficulty of activity on the estimation of time. *J. exp. Psychol.*, 1938, **23**, 270–287, 428–433.

HARTON, J. J. The influence of the degree of unity of organization. *J. gen. Psychol.*, 1939a, **21**, 25–49.

HARTON, J. J. An investigation of the influence of success and failure on the estimation of time. *J. gen. Psychol.*, 1939b, **21**, 51–62.

HARTON, J. J. The relation of time estimates to the actual time. *J. gen. Psychol.*, 1939c, **21**, 219–224.

HARTON, J. J. Time estimation in relation to goal organization and difficulty of tasks. *J. gen. Psychol.*, 1942, **27**, 63–69.

HAVET, J. *Kant et le problème du temps.* Paris: Gallimard, 1946.

HAWICKHORST, L. Mit welcher Sicherheit wird der Zeitwert einer Sekunde erkannt? *Z. Sinnesphysiol.*, 1934, **65**, 58–86.

HAWKES, G. R., BAILEY, R. W., & WARM, J. S. Method and modality in judgments of brief stimulus duration. *U. S. Army Med. Research Lab.*, *Fort Knox, Ky*, 1960, No. 422.

HEAD, H. *Studies in neurology.* London: Frowde, 1920.

HEBB, D. O. *The organization of behavior.* New York: Wiley, 1949.

HEIDEGGER, M. Sein und Zeit. *Jahrb. Phil. phänomenol. Forsch.*, 1927, **8**, 1–438.

HELM, W. Die Beeinflussung des Vergleichwertes einer Zeitstrecke durch ihre Verkoppelung mit einer zweiten teilweise gleichzeitigen Zeitstrecke. *Arch. ges. Psychol.*, 1937, **98**, 490–556.

HELSON, H., & KING, S. M. The tau effect. An example of psychological relativity. *J. exp. Psychol.*, 1931, **14**, 202–218.

HENRI, V. Analyse psychologique du principe de relativité. *J. Psychol. norm. path.*, 1920, **71**, 743–768.

HENRIKSON, E. H. A study of stage fright and the judgment of speaking time. *J. appl. Psychol.*, 1948, **32**, 532–536.

HENRY, F. M. Discrimination of the duration of a sound. *J. exp. Psychol.*, 1948, **38**, 734–743.

HERON, W. T. Time discrimination in rats. *J. comp. physiol. Psychol.*, 1949, **42**, 27–31.

HEYMANS, G., & WIERSMA, E. Beiträge zur speziellen Psychologie auf Grund einer Massenuntersuchung. *Z. Psychol., Physiol. Sinnesorg.*, 1909, **51**, 1–72.

HIEBEL, G., & KAYSER, CH. Le rythme nycthéméral de l'activité et de la calorification chez l'embryon de poulet et le jeune poulet. *C. R. Soc. Biol.*, 1949, **143**, 864–866.

HILGARD, E. R., & MARQUIS, D. G. *Conditioning and learning.* New York, London: D. Appleton Century Co., 1940.

HINDLE, H. M. Time estimates as a function of distance travelled and relative clarity of a goal. *J. Pers.*, 1951, **19**, 483–490.

HIRSCH, I. J., BILGER, R. C., & DEATHRAGE, B. H. The effect of auditory and visual background on apparent duration. *Amer. J. Psychol.*, 1956, **69**, 561–574.

HOAGLAND, H. The physiological control of judgments of duration: evidence for a chemical clock. *J. gen. Psychol.*, 1933, **9**, 267–287.

HOAGLAND, H. Temperature characteristics of the Berger rhythm in man. *Science*, 1936a, **83**, 84–85.

HOAGLAND, H. Electrical brain waves and temperature. *Science*, 1936b, **84**, 139–140.

HOAGLAND, H. Pacemakers of human brain waves. *Amer. J. Physiol.*, 1936c, **116**, 604–615.

HOAGLAND, H. Some pacemaker aspects of rhythmic activity in the nervous system. *Cold Spring Symposia on quantitative biology*, 1936d, **3**, 267–284.

HOAGLAND, H. The chemistry of time. *Sci. Mon.*, 1943, **56**, 56–61.

HOCHE, A. Langeweile. *Psychol. Forsch.*, 1923, **3**, 258–271.

HOFFMANN, K. Versuche zu der im Richtungsfinden der Vögel enthaltenen Zeitschätzung. *Z. Tierpsychol.*, 1954, **11**, 453–475.

HOFFMANN, K. Aktivitätsregistrierungen bei frisch geschlüpften Eidechsen. *Z. vergl. Physiol.*, 1955, **37**, 253–262.

HOLLINGWORTH, H. L. The inaccuracy of movement. *Arch. of Psychol.*, 1909, **11**, No. 13.

HORANYI-HECHST, B. Zeitbewusstsein und Schizophrenie. *Arch. Psychiat.*, 1943, **116**, 287–292.

HÖRING, A. *Versuche über das Unterscheidungsvermögen des Hörsinnes für Zeitgrössen.* Tübingen: 1864.

HORST, L. VAN DER. Le sens de la temporalisation pour la mémoire et pour l'orientation. *Encéphale*, 1956, 189–205.

HUANT, E., *Connaissance du temps.* Paris: Lethielleux, 1950.

HUBERT, H., & MAUSS, M. *Mélanges d'histoire des religions.* Paris: Alcan, 1909.

HULL, C. L. The goal gradient hypothesis and maze learning. *Psychol. Rev.*, 1932, 39, 25–43.

HULL, C. L. The alleged inhibition of delay in trace conditioned reactions. *Psychol. Bull.*, 1934, 31, 716–717.

HULL, C. L. The rat's speed of locomotion gradient in the approach to food. *J. comp. Psychol.*, 1934, 17, 393–422.

HULL, C. L. *Principles of Behavior.* New York: Appleton Century Co., 1943.

HULSER, C. Zeitauffassung und Zeitschätzung verschieden ausgefüllter Intervalle unter besonderen Berücksichtigung der Aufmerksamkeitsablenkung. *Arch. ges. Psychol.*, 1924, 49, 363–378.

HUME, D. *A treatise on human nature.* Ed. T. H. Green & T. H. Grose. 2 vols. London: Longmans, Green & Co., 1874.

HUNT, J. MCV., & SCHLOSBERG, H. General activity in the male white rat. *J. comp. Psychol.*, 1939a, 28, 23–38.

HUNT, J. MCV., & SCHLOSBERG, H. The influence of illumination upon general activity in normal, blinded and castrated male white rats. *J. comp. Psychol.*, 1939b, 28, 285–298.

HUNTER, W. S. Delayed reactions in animals and children. *Behav. Monogr.*, 1913, 2, No. 1.

HUSSERL, E. Vorlesungen zur Phänomenologie des inneren Zeitbewusstseins. *Jahrb. Phil. phänomenol. Forsch.*, 1928, 9, 367–496.

IRWIN, F. W., ARMITT, F. M., & SIMON, C. W. Studies in object preferences. I: The effect of temporal proximity. *J. exp. Psychol.*, 1943, 33, 64–72.

IRWIN, F. W., ORCHINIK, C. W., & WEISS, J. Studies in object preferences: the effect of temporal proximity upon adults' preferences. *Amer. J. Psychol.*, 1946, 59, 458–462.

ISRAELI, N. The psychopathology of time. *Psychol. Rev.*, 1932, 39, 486–491.

ISRAELI, N. Illusions in the perception of short time intervals. *Arch. of Psychol.*, 1930, 19, No. 113.

ISRAELI, N. Abnormal personality and time. New York: Science Printing, 1936.

ISRAELI, N. Ambiguous sound patterns: time of perception of variable non-visual figure ground and part whole relationships. *J. Psychol.*, 1940, 29, 449–452.

ISRAELI, N. The aesthetics of time. *J. gen. Psychol.*, 1951, 45, 259–263.

JACOBSEN, C. F. Studies of cerebral function in primates. *Comp. Psychol. Monogr.*, 1936, **13**(63), 1–60.

JAENSCH, E. R. Struktur-psychologische Erläuterung zur philosophischen Zeitlehre insbesondere bei Bergson und Proust. *Z. Psychol.*, 1932, **124**, 55-92.

JAENSCH, E. R., & KRETZ, A. Experimentelle strukturpsychologische Untersuchungen über die Auffassung der Zeit unter Berücksichtigung der Personaltypen. *Z. Psychol.*, 1932, **126**, 312–375.

JAHODA, M. Some socio-psychological problems of factory life. *Brit. J. Psychol.*, 1941, **31**, 191–206.

JAMES, W. *Principles of psychology.* 2 vols. London: Macmillan, 1891.

JAMES, W. *Psychology, briefer course.* New York: Holt, 1892.

JAMPOLSKY, P. Une nouvelle épreuve psychomotrice. *Rev. Psychol. appl.*, 1951, **1**, 103–138.

JANET, PAUL. Une illusion d'optique interne. *Rev. phil.*, 1877, **1**, 497–502.

JANET, PIERRE. *L'évolution de la mémoire et de la notion de temps.* Paris: Chahine, 1928.

JASPER, H., & SHAGASS, C. Conscious time judgments related to conditioned time intervals and voluntary control of the alpha rhythm. *J. exp. Psychol.*, 1941, **28**, 503–508.

JASPERS, K. *Allgemeine Psychopathologie.* Berlin: Julius Springer, 1920.

JENSEN, E. M., REESE, E. R., & REESE, T. W. The subitizing and counting of visually presented fields of dots. *J. Psychol.*, 1950, **30**, 363–392.

JONES, R. E. Personality changes in psychotics following prefrontal lobotomy. *J. abnorm. soc. Psychol.*, 1949, **44**, 315–328.

KALMUS, H. Über die Natur des Zeitgedächtnisses der Bienen. *Z. vergl. Physiol.*, 1934, **20**, 405–419.

KANT, E. *The critique of pure reason.* Trans. N. K. Smith. New York and London: Macmillan, 1934.

KASTENHOLZ, J. Untersuchungen zur Psychologie der Zeitauffassung. *Arch. ges. Psychol.*, 1922, **43**, 171–228.

KATZ, D. Experimentelle Beiträge zur Psychologie des Vergleichs im Gebiet des Zeitsinns. *Z. Psychol., Physiol. Sinnesorg.*, 1906, **42**, 302–340, 414–450.

KAWASIMA, S. The influence of time intervals upon the perception of arm motion. *Jap. J. Psychol.*, 1937, **12**, 270–289.

314 THE PSYCHOLOGY OF TIME

KAYSER, CH. Le rythme nycthéméral des mouvements d'énergie. *Rev. Scient.*, 1952, 90, 173–188.

KIESOW, F. Über die Vergleichung linearer Strecken und ihre Beziehung zum Weberschen Gesetze. *Arch. ges. Psychol.*, 1925, 52, 61–90; 53, 433–446.

KIMBLE, G. A. Conditioning as a function of the time between conditioned and unconditioned stimuli. *J. exp. Psychol.*, 1947, 37, 1–15.

KIRCHER, H. Die Abhängigkeit der Zeitschätzung von der Intensität des Reizes. *Arch. ges. Psychol.*, 1926, 54, 85–128.

KLEIN, L. Beitrag zur Psychopathologie und Psychologie des Zeitsinnes. *Z. Pathopsychol.*, 1917, 3.

KLEIST, K. *Gehirnpathologie.* Leipzig: Barth., 1934.

KLEITMAN, N. *Sleep and wakefulness as alternating phases in the cycle of existence.* Chicago: University of Chicago Press, 1939.

KLEITMAN, N., TITELBAUM, S., & HOFFMANN, H. The establishment of the diurnal temperature cycle. *Amer. J. Physiol.*, 1937, 119, 48–54.

KLEMM, O. Über die Wirksamkeit kleinster Zeitunterschiede. *Arch. ges. Psychol.*, 1925, 50, 204–220.

KLINEBERG, O. *Social psychology.* 2nd ed. New York: Holt, 1954.

KLINES, K., & MESZAROS, A. Der Rhythmus als biologisches Prinzip. Seine Genese und pathologische Bedeutung. *Arch. Psychiat. Nervenkrankh.*, 1942–43, 115, 90–112.

KLOOS, G. Störungen des Zeiterlebens in der endogenen Depression. *Nervenarzt*, 1938, 11, 225–244.

KOEHNLEIN, H. Über das absolute Zeitgedächtnis. *Z. Sinnesphysiol.*, 1934, 65 (1–2), 35–57.

KOFFKA, K. *Principles of Gestalt Psychology.* New York: Harcourt, 1935.

KOHLMANN, T. Das psychologische Problem der Zeitschätzung und der experimentelle Nachweis seiner diagnostischen Anwendbarkeit. *Wien. Z. Nervenheilk.*, 1950, 3, 241–260.

KOLLERT, J. Untersuchungen über den Zeitsinn. *Phil. Stud.*, 1883, 1, 78–89.

KONCZEWSKA, H. Les métamorphoses de l'espace et du temps dans la mémoire. *J. Psychol. norm. path.*, 1949, 42, 481–490.

KORNGOLD, S. Influence du genre de travail sur l'appréciation des grandeurs temporelles. *Travail hum.*, 1937, 1, 18–34.

KOTAKE, Y., & TAGWA, K. On the delay of the conditioned galvanic skin reflex in man. *Jap. J. Psychol.*, 1951, **22**, 1–6.

KOUPALOV, P. S. Periodic oscillations in the excitability of the cortex in a rhythmic alternation of positive reflexes and inhibitions. In *The work of Pavlov's physiological laboratories.* Vol. V. Moscow: 1954, 337–344. (In Russian.)

KOUPALOV, P. S., & PAVLOV, N. N. The action of the short conditioned stimulus in the case of delayed conditioned reflex. *Fiziol. Zh. S.S.S.R.*, 1935, **18**, 734–738. (In Russian.)

KOWALSKI, W. J. The effect of delay upon the duplication of short temporal intervals. *J. exp. Psychol.*, 1943, **33**, 239–246.

KRAMER, G. Experiments on bird orientation. *Ibis*, 1952, **94**, 265–285.

KURODA, R. The properties of time perception reproduced under muscular trace due to different quantities of weight. *Acta Psychol.*, *Keijo*, 1931, **1**, 83.

KURODA, R. Time estimation of longer intervals in the white rat. *Acta Psychol.*, *Keijo*, 1936, **2**, 155–159.

LA GARZA, C. O. DE, & WORCHEL, R. Time and space orientation in schizophrenics. *J. abnorm. soc. Psychol.*, 1956, **52**, 191–194.

LANGER, J., WAPNER, S., & WERNER, H. The effect of danger upon the experience of time. *Amer. J. Psychol.*, 1961, **74**, 94–97.

LAVELLE, L. *Du temps et de l'éternité.* Paris: Aubier, 1945.

LE BEAU, J. *Psycho-chirurgie et fonctions mentales.* Paris: Masson, 1954.

LECOMTE DU NOUY. *Le temps et la vie.* Paris: Gallimard, 1936.

LE GRAND, A. Recherches expérimentales sur la durée du rêve au moyen d'ingestions de bromure d'acétylcholine. *J. Physiol., Path. gén.*, 1949, **41**(2), 203.

LEMMON, V. W. The relation of reaction time to measures of intelligence, memory and learning. *Arch. of Psychol.*, 1927, **15**, No. 94.

LEONOW, W. A. Über die Bildung von bedingten Spurenreflexen bei Kindern. *Pflüg. Arch. ges. Physiol.*, 1926, **214**, 305–319.

LERIDON, S. & LE NY, J. F. Influence de l'inhibition de retard sur le temps de réaction motrice. *Bull. Centre Étud. Rech. psychotechn.*, 1955, **4**, 259–268.

LE SENNE, R. *Traité de caractérologie.* Paris: Presses Univ. de France, 1945.

LESHAN, L. L. Time orientation and social class. *J. abnorm. soc. Psychol.*, 1952, 47, 589–592.

LEVINE, M., & SPIVACK, G. Time conception and self control in a group of emotionally disturbed adolescents. *J. clin. Psychol.*, 1959, 15, 224–226.

LEWIN, K. *A dynamic theory of personality.* New York: McGraw-Hill, 1935.

LEWIN, K. Behavior and development as a function of the total situation. In L. CARMICHAEL, *Manual of child psychology.* New York: Wiley, 1946.

LEWIS, A. Experience of time in mental disorder. *Proc. Roy. Soc. Med.*, 1932, 25, 611–620.

LEWIS, M. M. The beginning of reference to past and future in a child's speech. *Brit. J. educ. Psychol.*, 1937, 7, 39–56.

LIPPS, TH. *Grundtatsachen des Seelenlebens.* Bonn: Cohen, 1883.

LOCKE, J. *Essay concerning human understanding.* Ed. Alexander Campbell Fraser, 2 vols. Oxford: Clarendon Press, 1894.

LOEHLIN, J. C. The influence of different activities on the apparent length of time. *Psychol. Monog.*, 1959, 73, No. 474.

LOOMIS, E. A., JR. Space and time perception and distortion in hypnotic states. *Personality*, 1951, 1, 283–293.

LOSSAGK, H. Experimenteller Beitrag zur Frage des Monotonie-Empfindens. *Industr. Psychotechnik*, 1930, 7, 101–107.

LOTZE, H. *Medicinische Psychologie.* Leipzig: Weidmann, 1952.

MAACK, A. Untersuchungen über die Anwendbarkeit des Weber-Fechnerschen Gesetzes auf die Variation der Lautdauer. *Z. Phonet.*, 1948, 2, 1–15.

MACH, E. Untersuchungen über den Zeitsinn des Ohres. *Sitz. Wien. Akad. Wiss.*, 1865, Kl., 51.

MAGER, A. Neue Versuche zur Messung der Geschwindigkeit der Aufmerksamkeitswanderung. *Arch. ges. Psychol.*, 1925, 53, 391–432.

MALMO, R. B. Interference in delayed response in monkeys after removal of the frontal lobes. *J. Neurophysiol.*, 1942, 5, 295–308.

MALRIEU, PH. *Les origines de la conscience du temps.* Paris: Presses Univ. de France, 1953.

MARQUIS, D. P. Learning in the neonate. *J. exp. Psychol.*, 1941, 29, 263–282.

MARROU, H. I. L'ambivalence du temps de l'histoire chez saint Augustin. Paris: Vrin, 1950.

MARTIN, L. La mémoire chez convoluta. Thesis, faculty of science, Paris, 1900.

MARX, CH., & KAYSER, CH. Le rythme nycthéméral de l'activité chez le lézard. C. R. Soc. Biol., 1949, 143, 1375–1377.

MAURY, A. Le sommeil et les rêves. Études psychologiques. Paris: Didier, 1861.

MAYO, R. Is there a sense of duration? Mind, 1950, 59, 71–78.

MCALLISTER, W. R. Eyelid conditioning as a function of the CS-US interval. J. exp. Psychol., 1953a, 45, 417–422.

MCALLISTER, W. R. The effect on eyelid conditioning of shifting the CS-US interval. J. exp. Psychol., 1953b, 45, 423–428.

MCDOUGALL, R. Sex differences in the sense of time. Science, 1904, 19, 707–708.

MCGRILL, V. J. An analysis of the experience of time. J. Phil., 1930, 27, 533–544.

MCLEOD, R. B., & ROFF, M. F. An experiment in temporal disorientation. Acta psychol., 1938, 1, 381–423.

MEDIONI, J. L'orientation "astronomique" des arthropodes et des oiseaux. Ann. Biol., 1956, 32, 37–67.

MEERLOO, A. M. Father time: an analysis of subjective conceptions of time. Psychiat. Quart., 1948, 22, 587–608.

MEHNER, M. Zur Lehre vom Zeitsinn. Phil. Stud., 1885, 2, 546–602.

MERLEAU-PONTY, M. Phénoménologie de la perception. Paris: Gallimard, 1945.

MERLEAU-PONTY, M. Time consciousness in Husserl and Heidegger. Phil. Phenomenol. Res., 1947, 8, 23–54.

METZ, B. et al. Le rythme nycthéméral de la température et de la calorification. J. Physiol., Path. gén., 1952, 44, 135–142.

MEUMANN, E. Beiträge zur Psychologie des Zeitsinns. Phil. Stud., 1893, 8, 431–519; 1894a, 9, 264–306.

MEUMANN, E. Untersuchungen zur Psychologie und Aesthetik des Rhythmus. Phil. Stud., 1894b, 10, 249–322, 393–430.

MEUMANN, E. Beiträge zur Psychologie des Zeitbewusstseins. Phil. Stud., 1896, 12, 128–254.

MICHAUD, E. *Essai sur l'organisation de la connaissance entre 10 et 14 ans.* Paris: Vrin, 1949.

MICHOTTE, A. La simultanéité apparente. *Ann. Inst. Sup. Phil.*, 1912, **1**, 568–663.

MILLER, G. A., & TAYLOR, W. G. The perception of repeated bursts of noise. *J. acoust. Soc. Amer.*, 1948, **20**, 171–182.

MILLER, G. A., & LICKLIDER, J. C. R. Intelligibility of interrupted speech. *J. acoust. Soc. Amer.*, 1950, **22**, 167–173.

MILLER, N. E. Experimental studies in conflict. In J. MCV. HUNT, *Personality and the behavior disorders.* Vol. I. New York: Ronald Press, 1944, pp. 431–465.

MINKOWSKI, E. Le problème du temps en psychopathologie. *Rech. phil.*, 1932–33, **2**, 231–256.

MINKOWSKI, E. *Le temps vécu.* In the series *L'Évolution Psychiatrique.* Paris: Centre d'Edit. Psychi., 1933.

MINKOWSKI, E. Le problème du temps chez Pierre Janet. *Évol. psychiat.*, No. 3, 1950, 451–463.

MONASTERIO, R. I. Tiempo y ritmo in psicologia. *Riv. Psicol. gen. apl.*, 1947, **2**, 479–485.

MORAND. Le problème de l'attente. *Année psychol.*, 1914–19, **21**, 1 78.

MOREAU DE TOURS, J. *Du haschisch et de l'aliénation mentale.* Paris: Fortin, Masson, 1845.

MORGAN, C. T., STELLAR, E., & JOHNSON, O. Food-deprivation and hoarding in rats. *J. comp. Psychol.*, 1943, **35**, 275–295.

MORI, T. Experimental studies of time discrimination in the white rat. *Annu. anim. Psychol.*, 1954, **4**, 7–16.

MOWBRAY, G. H., & GEBHARD, J. W. The differential sensitivity of the eye to intermittence. *Amer. Psychologist*, 1954, **9**, 436.

MOWRER, O. H. *Learning theory and personality dynamics.* New York: Ronald Press, 1950.

MOWRER, O. H., & LAMOREAUX, R. R. Avoidance conditioning and signal duration—a study of secondary motivation and reward. *Psychol. Monogr.*, 1942, **54**, No. 5.

MUNDLE, C. W. K. How specious is the "specious present"? *Mind*, 1954, **63**, 26–48.

MÜNSTERBERG, J. *Beiträge zur experimentellen Psychologie.* Heft 2. Freiburg-Br.: Siebeck, 1889.

MUSATTI, C. L. *Elementi di psicologia della testimonianza.* Padua: Cedam, 1931.

MYERS, G. C. Incidental perception. *J. exp. Psychol.*, 1916, 1, 339–350.

NELSON, M. L. The effect of sub-divisions on the visual estimate of time. *Psychol. Rev.*, 1902, 9, 447–459.

NEULAT, G. *Contribution à l'étude du travail de nuit.* Thesis, faculty of medicine, Lyons, 1950.

NEUMANN, N. Die Konkurrenz zwischen den Auffassungen der Zeitdauer und deren Ausfüllung bei verschiedener Einstellung der Aufmerksamkeit. *Arch. ges. Psychol.*, 1936, 95, 200–255.

NICHOLS, H. The psychology of time. *Amer. J. Psychol.*, 1890, 3, 453–529.

NITARD, Y. Apparent time acceleration with age. *Science*, 1943, 98, No. 2535.

NOGUÉ, J. Ordre et durée. *Rev. phil.*, 1932, 64, 45–76.

NOGUÉ, J. Le système de l'actualité. *J. Psychol. norm. path.*, 1939, 36, 344–369.

OACKDEN, E. C., & STURT, M. The development of the knowledge of time in children. *Brit. J. Psychol.*, 1922, 12, 309–337.

OBERNDORF, C. P. Time. Its relation to reality and purpose. *Psychoanal. Rev.*, 1941, 28, 139–155.

ODIER, CH. Le réveille-matin diencéphalique. *Rev. suisse psychol.*, 1946, 5, 113–117.

OLÉRON, G. Influence de l'intensité d'un son sur l'estimation de sa durée apparente. *Année psychol.*, 1952, 52, 383–392.

OMBREDANE, A. *L'aphasie et l'élaboration de la pensée explicite.* Paris: Presses Univ. de France, 1951.

OMWAKE, K. T., & LORANZ, M. Study of ability to wake at a specified time. *J. appl. Psychol.*, 1933, 17, 468–474.

OSBORNE, A. Body temperature and periodicity. *J. Physiol.*, 1907, 36, 39.

PAILLARD, J. Quelques données psychophysiologiques relatives au déclenchement de la commande motrice. *Année psychol.*, 1947–1948, 46–47, 26–47.

PALIARD, J. Le temps. *Étud. phil.*, 1954, 9, 393–399.

PASCAL, B. De l'esprit géometrique. In *Pensées et opuscules.* Ed. Brunschvicg. Paris: Hachette, 1909.

PAULHAN, F. Le présentisme. *Rev. phil.*, 1924, 49, 190–237.

PAULHAN, F. L'influence psychologique et les associations du présentisme. I: Les traits de caractère subordonnés du présentisme; II:

Quelques groupes de présentistes. *J. Psychol. norm. path.*, 1925, **22**, 193–235, 297–325.

PAVLOV, I. P. *Conditioned reflexes; an investigation of the physiological activity of the cerebral cortex* (English translation). London: Oxford University Press, 1927.

PAVLOV, I. P. *Lectures on conditioned reflexes; 25 years of objective study of the higher nervous activity of animals* (English translation). New York: International, 1928.

PERES, J. Causes d'inegalité d'évaluation de la durée. *J. Psychol. norm. path.*, 1909, **6**, 227–231.

PETERS, R. H., ROSVOLD, H. E., & MIRSKY, A. F. The effect of thalamic lesions upon delayed response-type tests in the Rhesus monkey. *J. comp. physiol. Psychol.*, 1956, **49**, 111–116.

PETRIE, A. *Personality and the frontal lobes.* London: Routledge & Kegan Paul, 1952.

PHILIP, B. R. The anchoring of absolute judgments of short temporal intervals. *Bull. Canad. Psychol. Ass.*, 1944, **4**, 25–28.

PHILIP, B. R. The effect of interpolated and extrapolated stimuli on the time order error in the comparison of temporal intervals. *J. gen. Psychol.*, 1947, **36**, 173–187.

PIAGET, J. *La construction du réel chez l'enfant.* Neuchâtel, Paris: Delachaux & Niestlé, 1937.

PIAGET, J. *Les notions de mouvement et la vitesse chez l'enfant.* Paris: Presses Univ. de France, 1946a.

PIAGET, J. *Le développement de la notion de temps chez l'enfant.* Paris: Presses Univ. de France, 1946b.

PIAGET, J. *Épistémologie génétique.* 3 vols. Paris: Presses Univ. de France, 1950.

PICHON, E. Essai d'étude convergente des problèmes du temps. *J. Psychol. norm. path.*, 1931, **28**, 85–118.

PICK, A. Psychopathologie des Zeitsinns. *Z. Pathopsychol.*, 119, **3**, 430–441.

PIÉRON, H., & VASCHIDE, N. La valeur séméiologique du rêve. *Rev. scient.*, 1901, **15**, 385–399, 427–430.

PIÉRON, H. *L'évolution de la mémoire.* Paris: Flammarion, 1910.

PIÉRON, H. *Le problème physiologique du sommeil.* Paris: Masson et Cie, 1913.

PIÉRON, H. Les problèmes psychophysiologiques de la perception du temps. *Année psychol.*, 1923, **24**, 1–25.

PIÉRON, H. La persistance à l'obscurité du rythme lumineux du lampyre. *Feuille nat.*, 1925, **21**, 186–188.

PIÉRON, H. L'attention. In G. DUMAS, *Nouveau traité de psychologie*. Vol. IV. Paris: Alcan, 1934.

PIÉRON, H. L'évanouissement de la sensation lumineuse. *Année psychol.*, 1934, **35**, 1–49.

PIÉRON, H. Quelques réflexions et observations à propos de l'induction des rythmes chez les animaux. *J. Psychol. norm. path.*, 1937, **34**, 397–412.

PIÉRON, H. Psychologie zoologique. In G. DUMAS, *Nouveau traité de psychologie*. Vol. VIII. Paris: Presses Univ. de France, 1941.

PIÉRON, H. Le problème du temps au point de vue de la psychophysiologie. *Sciences*, 1945, **72**, 28–41.

PIÉRON, H. Des aspects réels du temps en psychophysiologie. In *Essays in Psychology dedicated to D. Katz*. Uppsala: Almquist & Wiksells, 1951, pp. 214–222.

PIÉRON, H. *La sensation guide de vie*. 3rd ed. Paris: Gallimard, 1955.

PINTNER, R. Standardization of the Knox cube test. *Psychol. Rev.*, 1915, **22**, 377–401.

PISTOR, F. Measuring the time concepts of children. *J. educ. Res.*, 1939, **33**, 293–300.

PITTENDRICH, C. S. On temperature independence in the clock system controlling emergence time in drosophila. *Proc. Nat. Acad. Sci.*, Wash., 1954, **40**, 1018–1029.

POINCARÉ, H. *La valeur de la science*. Rev. ed. Paris: Flammarion, n.d.

POIRIER, R. Temps spirituel et temps matériel. *Rech. phil.*, 1935–36, **5**, 1–40.

POPOV, N. A. Le facteur temps dans la théorie des réflexes conditionnés. *C. R. Soc. Biol.*, 1948, **142**, 156–158.

POPOV, N. A. Action prolongée sur le cortex cérébral après stimulation rythmique. *J. Physiol. Path. gén.*, 1950a, **42**, 51–72.

POPOV, N. A. *Études de psychophysiologie*. Paris: Les Éditions du Cèdre, 1950b.

POROT, M. La leucotomie préfrontale en psychiatrie. *Ann. méd. psychol.*, 1947, **105**, 121–142.

POSTMAN, L. Estimates of time during a series of tasks. *Amer. J. Psychol.*, 1944, **57**, 421–424.

POSTMAN, L., & MILLER, G. A. Anchoring of temporal judgments. *Amer. J. Psychol.*, 1945, **58**, 42–53.

322 THE PSYCHOLOGY OF TIME

POULET, G. *Études sur le temps humain.* Paris: Plon, 1950.

PUCELLE, J. *Le temps.* Paris: Presses Univ. de France, 1955.

PUMPIAN-MINDLIN, E. Über die Bestimmung der bewussten Zeitschätzung bei normalen und dementen Epileptikern. *Arch. suiss. Neurol.*, 1935, 36, 291–305.

QUASEBARTH, K. Zeitschätzung und Zeitauffassung optisch und akustisch ausgefüllter Intervalle. *Arch. ges. Psychol.*, 1924, 49, 379–432.

RANSCHBURG, P. Les bases somatiques de la mémoire. In *Centenaire de Th. Ribot.* Paris: Imprimerie Moderne, Agen., 1939.

REGELSBERGER, H. Über die cerebrale Beeinflussung der vegetativen Nahrungsrythmik. *Z. ges. Neurol. Psychiat.*, 1940, 169, 532–542.

RÉGIS, E. *Précis de Psychiatrie.* 6th ed. Paris: Doin, 1923.

REGNAUD, P. L'idée de temps, origine des principales expressions qui s'y rapportent dans les langues indo-européennes. *Rev. phil.*, 1885, 19, 280–287.

REICHLE, F. Untersuchungen über Frequenzrhythmen bei Ameisen. *Z. vergl. Physiol.*, 1943, 30, 227–251.

REMLER, O. Untersuchungen an Blinden über die 24-Stunden Rhythmik. *Klin. Mbl. Augenheilk.*, 1949, 113, 116–137.

RENNER, M. Ein Transozeanversuch zum Zeitsinn der Honigbiene. *Naturwiss.*, 1955, 42, 540–541.

RENSHAW, S. An experimental comparison of the production and auditory discrimination by absolute impression of a constant tempo. *Psychol. Bull.*, 1932, 29(9), 659.

REVAULT D'ALLONES, G. Rôle des sensations internes dans les émotions et la perception de la durée. *Rev. phil.*, 1905, 2, 592–623.

RIBOT, TH. *La psychologie allemande contemporaine.* Paris: G. Baillière, 1879.

RICHET, CH. Forme et durée de la vibration nerveuse et l'unité psychologique de temps. *Rev. phil.*, 1898, 45, 337.

RIGBY, W. K. Approach and avoidance gradients and conflict behavior in a predominantly temporal situation. *J. comp. physiol. Psychol.*, 1954, 47, 83–89.

RIZENDE, M. DE. Uma experiencia sôbre a percepçâo do tempo. *Arq. Brasil. Psicotécnica*, 1950, 2, 40–55.

ROBERTS, W. H. The effect of delayed feeding on white rats in a problem cage. *J. genet. Psychol.*, 1930, 37, 35–38.

RODNICK, E. H. Characteristics of delayed and trace conditioned responses. *J. exp. Psychol.*, 1937a, **20**, 409–425.

RODNICK, E. H. Does the interval of delay of conditioned responses possess inhibitory properties? *J. exp. Psychol.*, 1937b, **20**, 507–527.

ROELOFS, O., & ZEEMAN, W. P. G. The subjective duration of time intervals. *Acta Psychol.*, 1949, **6**, 126–177, 289–336.

ROKEACH, M. The effect of perception time upon rigidity and concreteness of thinking. *J. exp. Psychol.*, 1950, **40**, 206.

ROSENBAUM, G. Temporal gradients of response strength with two levels of motivation. *J. exp. Psychol.*, 1951, **41**, 261–267.

ROSENBERG, M. Über Störungen der Zeitschätzung. *Z. ges. Neurol. Originalien. Psychiat.*, 1919, **51**, 208–223.

ROSENZWEIG, S. Preferences in the repetition of successful and unsuccessful activities as a function of age and personality. *J. genet. Psychol.*, 1933, **42**, 423–441.

ROSENZWEIG, S., & KOHT, A. G. The experience of duration as affected by need tension. *J. exp. Psychol.*, 1933, **16**, 745–774.

ROSS, S., & FLETCHER, J. L. Response time as an indicator of color deficiency. *J. appl. Psychol.*, 1953, **37**, 211–214.

ROSS, S., & KATCHMAR, L. The construction of a magnitude function for short-time intervals. *Amer. J. Psychol.*, 1951, **64**, 397–401.

RUBIN, E. Geräuschverschiebungsversuche. *Acta Psychol.*, 1932, **4**, 203–236.

RUBIN, E. Some elementary time experiences. *Acta Psychol.*, 1935, **1**, 206–211.

RUCH, F. L. L'appréciation du temps chez le rat blanc. *Année psychol.*, 1931, **32**, 118–130.

SAMS, C. F., & TOLMAN, E. C. Time discrimination in white rats. *J. comp. Psychol.*, 1925, **5**, 255–263.

SAUTER, V. Versuche zur Frage des "Zähl" Vermögens bei Elstern. *Z. Tierpsychol.*, 1952, **9**, 252–289.

SCHAEFER, G., & GILLILAND, R. The relation of time estimation to certain physiological changes. *J. exp. Psychol.*, 1938, **23**, 545–552.

SCHEEVOIGT, W. Die Wahrnehmung der Zeit bei den verschiedenen Menschentypen. *Z. Psychol.*, 1934, **131**, 217–295.

SCHILDER, P. Psychopathology of time. *J. nerv. ment. Dis.*, 1936, **83**, 530–546.

SCHNEIDER, L., & LYSGAARD, S. The deferred gratification pattern: a preliminary study. *Social Forces*, 1953, **18**, 142–149.

SCHULTZE, O. Beiträge zur Psychologie des Zeitbewusstseins. *Arch. ges. Psychol.*, 1908, **13**, 275–351.

SCHUMANN, F. Zur Psychologie der Zeitanschauung. *Z. Psychol. Physiol. Sinnesorg.*, 1898, **17**, 106–148.

SCOTT, W. C. N. Some psycho-dynamic aspects of disturbed perception of time. *Brit. J. med. Psychol.*, 1948, **21**, 111–120.

SHERRINGTON, C. S. *The integrative action of the nervous system*. London: Constable, 1906.

SIVADJIAN, J. *Le temps*. Paris: Hermann, 1938.

SKALET, M. The significance of delayed reactions in young children. *Comp. Psychol. Monogr.*, 1930–31, **7**, No. 4.

SMITH, P. C. The prediction of individual differences in susceptibility to industrial monotony. *J. appl. Psychol.*, 1955, **39**, 322–329.

SOURIAU, M. *Le temps*. Paris: Alcan, 1937.

SPENCER, L. T. Experiments in time estimation using different interpolations. *Amer. J. Psychol.*, 1921, **32**, 557–562.

SPIEGEL, E. A., WYCIS, H. T., ORCHINIK, C. W., & FREED, H. The thalamus and temporal orientation. *Science*, 1955, **121**, 770–771.

SPOONER, A., & KELLOG, W. N. The backward conditioning curve. *Amer. J. Psychol.*, 1947, **60**, 321–334.

SPRINGER, D. Development in young children of an understanding of time and the clock. *J. genet. Psychol.*, 1952, **80**, 83–96.

STEIN, H. Untersuchungen über den Zeitsinn der Vögel. *Z. vergl. Physiol.*, 1951, **33**, 387–403.

STEIN-BELING, VON. Über das Zeitgedächtnis bei Tieren. *Biol. Rev.*, 1935, **10**, 18.

STEINBERG, A. Changes in time perception induced by an anaesthetic drug. *Brit. J. Psychol.*, 1955, **46**, 273–279.

STERN, L. W. Psychische Präsenzzeit. *Z. Psychol. Physiol. Sinnesorg.*, 1897, **13**, 325–349.

STERN, L. W., STERN, CL. *Die Sprache des Kindes*. Leipzig: Barth, 1907.

STERZINGER, O. Chemopsychologische Untersuchungen über den Zeitsinn. *Z. Psychol.*, 1935, **134**, 100–131.

STERZINGER, O. Neue chemopsychologische Untersuchungen über den menschlichen Zeitsinn. *Z. Psychol.*, 1938, **143**, 391–406.

STOETZEL, J. La pression temporelle. *Sondages*, 1953, **15**, 11–23.

STONE, S. A. Prior entry in the auditory-tactual complication. *Amer. J. Psychol.*, 1926, **37**, 284–287.

STOTT, L. H. The discrimination of short tonal durations. Ph.D. dissertation, University of Illinois, 1933.

STOTT, L. H. Time order errors in the discrimination of short tonal durations. *J. exp. Psychol.*, 1935, **18**, 741–766.

STRAUS, E. Das Zeiterlebnis in der endogenen Depression und in der psychopathischen Verstimmung. *Mschr. Psychiat. Neurol.*, 1928, **68**, 640–657.

STROUD, J. M. The fine structure of psychological time. In *Information theory in psychology*. Glencoe, Ill.: Free Press, 1956.

STURT, M. Experiments on the estimation of duration. *Brit. J. Psychol.*, 1923, **13**, 382–388.

STURT, M. The psychology of time. London: Kegan Paul, 1925.

SUDO, Y. On the effect of the phenomenal distance upon time perception. *Jap. J. Psychol.*, 1941, **16**, 95–115.

SUTO, Y. The effect of space on time estimation in tactual space. *Jap. J. Psychol.*, 1952, **22**, 189–201.

SUTO, Y. The effect of space on time estimation (S effect) in tactual space. II: The role of vision in the S effect upon the skin. *Jap. J. Psychol.*, 1955, **26**, 94–99.

SUTO, Y. Role of apparent distance in time perception. *Research rep., Tokyo Elect. Engineering Coll.*, 1959, **5**, 73–82.

SWEET, A. L. Temporal discrimination by the human eye. *Amer. J. Psychol.*, 1953, **66**, 185–198.

SWIFT, E. Y., & MCGEOCH, J. A. An experimental study of the perception of filled and empty time. *J. exp. Psychol.*, 1925, **8**, 240–249.

SWITZER, S. A. Anticipatory and inhibitory characteristics of delayed conditioned reactions. *J. exp. Psychol.*, 1934, **17**, 603–620.

SZYMANSKI, J. S. Die Haupttiertypen in Bezug auf die Verteilung der Ruhe- und Aktivitäts-perioden im 24-stündigen Zyklus. *Biol. Zbl.*, 1916, **36**, 357.

TAUBMAN, R. E. Studies in judged number. I: The judgment of auditory number. II: The judgment of visual number. *J. gen. Psychol.*, 1950, **43**, 167–194, 195–219.

TEAHAN, J. E. Future time perspective optimism and academic achievement. *J. abnorm. soc. Psychol.*, 1958, **57**, 379–380.

TEUBER, H. L., & BENDER, M. B. Alterations in pattern vision following trauma of occipital lobes in man. *J. gen. Psychol.*, 1949, **40**, 37–57.

THURY, M. L'appréciation du temps. *Arch. de Psychol.*, 1903, **2**, 182–184.

TINKER, M. A. Temporal perception. In BORING, E. G., LANGFELD, H. S., & WELD, H. P. *Psychology*. New York: Wiley, 1935.

TITCHENER, E. B. *Lectures on the elementary psychology of feeling and attention*. New York: Macmillan, 1908.

TOBOLOWSKA, J. *Étude sur les illusions de temps dans le rêve du sommeil normal*. Thesis, faculty of medicine, Paris, 1900.

TOULOUSE, E., & PIÉRON, H. Le mécanisme de l'inversion chez l'homme du rythme nycthéméral de la température. *J. Physiol. Path. gen.*, 1907, **3**, 425–440.

TRIPLETT, D. The relation between the physical pattern and the reproduction of short temporal intervals: a study in the perception of filled and unfilled time. *Psychol. Monogr.* 1931, **41**, 4(187), 201–265.

VASCHIDE, N. *Le sommeil et les rêves*. Paris: Flammarion, 1911.

VERLAINE, L. L'instinct et l'intelligence chez les hymenoptères. IX: La notion du temps. *Ann. Bull. Soc. Entom. Belg.*, 1929, **69**, 115–125.

VIERORDT, K. *Der Zeitsinn nach Versuchen*. Tübingen: H. Laupp, 1868.

VINCE, M. A. The intermittency of control movements and the psychological refractory period. *Brit. J. Psychol.*, 1948, **38**, 149–157.

VINCHON, J. Quelques exemples d'évaluation du temps chez les schizophrènes. *J. Psychol. norm. path.*, 1920, **17**, 415–417.

VINCHON, J., & MONESTIER. Nouvel exemple d'évaluation du temps par un schizophrène. *J. Psychol. norm. path.*, 1922, **17**, 735–738.

VISHER, A. L. Psychological problems of the aging personality. *Bull. schweiz. Akad. Wiss.*, 1947, **2**, 280–286.

VITELES, M. S. Le problème de l'ennui. *Travail hum.*, 1952, **15**, 85–98.

VOLMAT, R. *L'art psychopathologique*. Paris: Presses Univ. de France, 1955.

WAALS, H. G. VAN DER, & ROELOFS, C. O. Contenu de la perception et durée apparente de la perception. *Ned. Tijdschr. Psychol.*, 1946, **1**, 45–70; **2**, 150–204.

WAHL, O. Neue Untersuchungen über das Zeitgedächtnis der Bienen. *Z. vergl. Physiol.*, 1932, **16**, 529–589.

WAHL, O. Beitrag zur Frage der biologischen Bedeutung des Zeitgedächtnisses der Bienen. *Z. vergl. Physiol.*, 1933, **18**, 709.

WALLACE, M. Future time perspective in schizophrenia. *J. abnorm. soc. Psychol.*, 1956, **52**, 240–245.

WALLACE, M., & RABIN, A. I. Temporal experience. *Psychol. Bull.*, 1960, **57**, 213–236.

WALLON, H. *Les origines de la pensée chez l'enfant.* 2 vols. Paris: Presses Univ. de France, 1947.

WALLON, H. Le problème biologique de la conscience. In G. DUMAS, *Nouveau traité de psychologie.* Vol. I. Paris: Alcan, 1930.

WEBER, A. O. Estimation of time. *Psychol. Bull.*, 1933, **30**, 233–252.

WEBER, C. O. The properties of space and time in kinaesthetic field of force. *Amer. J. Psychol.*, 1926, **38**, 597–606.

WECHSLER, D. A study of retention in Korsakoff psychosis. *Psychiat. Bull.*, 1917, 1–49.

WEINSTEIN, A. D., GOLDSTONE, S., & BOARDMAN, W. K. The effect of recent and remote frames of reference on temporal judgments of schizophrenic patients. *J. abnorm. soc. Psychol.*, 1958, **57**, 241–244.

WELFORD, A. T. The "psychological refractory period" and the timing of high speed performance, a review and a theory. *Brit. J. Psychol.*, 1952, **43**, 2–20.

WERNER, H., & THUMA, B. D. A deficiency in the perception of apparent motion in children with brain injury. *Amer. J. Psychol.*, 1942, **55**, 58–67.

WHIPPLE, G. M. On nearly simultaneous clicks and flashes. *Amer. J. Psychol.*, 1898, **10**, 280–286.

WHITE, C. T., & SCHLOSBERG, H. Degree of conditioning of the GSR as a function of the period of delay. *J. exp. Psychol.*, 1952, **43**, 357–362.

WILSON, M. P., & KELLER, F. S. On the selective reinforcement of spaced responses. *J. comp. physiol. Psychol.*, 1953, **46**, 190–193.

WIRTH, W. Die unmittelbare Teilung einer gegebenen Zeitstrecke. *Amer. J. Psychol.*, 1937, **50**, 79–96.

WOLFLE, H. M. Time factors in conditioning finger-withdrawal. *J. gen. Psychol.*, 1930, **4**, 372–378.

WOLFLE, H. M. Conditioning as a function of the time interval between the unconditioned and conditioned stimulus. *J. gen. Psychol.*, 1932, **7**, 80–103.

WOODROW, H. Behavior with respect to short temporal stimulus forms. *J. exp. Psychol.*, 1928a, **11**, 167–193, 259–280.

WOODROW, H. Temporal discrimination in the monkey. *J. comp. Psychol.*, 1928*b*, 8, 395–427.

WOODROW, H. Discrimination by the monkey of temporal sequences of varying numbers of stimuli. *J. comp. Psychol.*, 1929, 9, 123–158.

WOODROW, H. The reproduction of temporal intervals. *J. exp. Psychol.*, 1930, 13, 473–499.

WOODROW, H. Individual differences in the reproduction of temporal intervals. *Amer. J. Psychol.*, 1933, 45, 271–281.

WOODROW, H. The temporal indifference interval determined by the method of mean error. *J. exp. Psychol.*, 1934, 17, 167–188.

WOODROW, H. The effect of practice upon time order errors in the comparison of temporal intervals. *Psychol. Rev.*, 1935, 72, 127–152.

WOODROW, H. Time perception. In S. S. STEVENS, *Handbook of experimental psychology.* New York: Wiley, 1951.

WOODWORTH, R. S. *Experimental psychology.* New York: Holt, 1938.

WUNDT, W. *Éléments de psychologie physiologique.* 2 vols. French trans. Rouvier. Paris: Alcan, 1886.

YERKES, R. M., & URBAN. Time estimation in its relation to sex, age and physiological rhythms. *Harv. Psychol. Stud.*, 1906, 2, 405–430.

‖‖‖‖‖‖‖‖‖‖‖‖‖‖‖‖

INDEXES

‖‖‖‖‖‖‖‖‖‖‖‖‖‖‖‖

INDEX OF AUTHORS

INDEX OF SUBJECTS